The Ocean World of Jacques Cousteau

Outer and Inner Space

The Ocean World of Jacques Cousteau

Outer and Inner Space

WORLD PUBLISHING

TIMES MIRROR
NEW YORK

The ocean world does not relinquish its secrets with ease but demands dilligence and daring, a sincere dedication to science, and an unflagging fascination for this last frontier.

Published by The World Publishing Company

Published simultaneously in Canada
by Nelson, Foster & Scott Ltd.

First Printing—1974

ISBN 0-529-05160-5
Library of Congress catalog card number: 73-21593

Printed in the United States of America

Project Director: Steven Schepp

Managing Editor: Richard C. Murphy

Assistant Managing Editor: Christine Names
Senior Editor: Ralph Slayton
Assistant to the Senior Editor: Robert Schreiber
Editorial Assistant: Joanne Cozzi

Art Director and Designer: Gail Ash

Assistant to the Art Director: Martina Franz
Illustrations Editor: Howard Koslow

Vice President, Production: Paul Constantini

Creative Consultant: Milton Charles

Typography: Nu-Type Service, Inc.

WORLD PUBLISHING
TIMES MIRROR
NEW YORK

Contents

From the time of the first great civilizations until quite recently, outer space—the heavens—claimed more of man's interest than inner space—the sea. Today, OUTER AND INNER SPACE claim equal attention, and we know that the frontier we had for so long ignored is crucial to our survival.

The earliest records we possess of what were to become the marine sciences are from about 1000 B.C. They are to be found in the works of the great poets like Homer and Hesiod, in tales of Ulysses' wanderings over the Mediterranean or in descriptions of lands beyond the Inland Sea. AN INFANT SCIENCE (Chapter I) might be said to have begun with them.

For many hundreds of years few advances were made, but with the Renaissance there was a great revival of curiosity, a rediscovery of ancient learning, and a renewed zeal for knowledge. The PIONEERS IN OCEAN SCIENCE (Chapter II) were able to draw on important scientific discoveries made in the seventeenth and eighteenth centuries and bring them to bear on their investigations.

In the twentieth century, equipment and techniques have improved so rapidly that the store of information about the sea has begun to grow at an astonishing rate. RECENT VOYAGES OF DISCOVERY (Chapter III) have brought together scientists from many nations, making for a broader sharing of knowledge.

Of the many sciences that help us know the sea, the one that has held the greatest fascination is the study of marine life. Recently, the biologist of the sea has had two very pragmatic concerns—one, the effects of pollution on marine life and, the other, the ways in which these organisms might be nurtured and better harvested to feed a great many hungry people. HOW MUCH LIFE IN THE SEA? (Chapter IV), a question man had long asked, had come to be of crucial importance.

Another question he asked may be how he might contend with THE DESTRUCTIVE FORCES OF THE SEA (Chapter V). Ever since man first built a home or a harbor at its edge, he has known the sea's power only too well. By subtle means or spectacular, waves and sand and tiny organisms worked to bring his

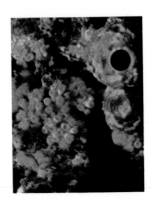

efforts to nothing. Through a persistent seeking after knowledge, we will understand those destructive forces.

The oceans are THE CLIMATIC REGULATOR (Chapter VI) for the whole globe, and we need a better understanding of the way in which that regulator works.

Considerable activity in the sea is directed toward the LOCATION OF MINERALS (Chapter VII). As a result of movements of the earth's crust some valuable minerals are released into the seawater, forming rich pools on ocean basins or precipitating out in nodules on the bottom.

The single most useful tool that we possess for the study of the sea is underwater acoustics. With sound we are able to measure many of the sea's physical properties, to delineate the topography of the bottom and of the sub-bottom structures, and to calculate the densities of marine populations. VIBRATIONS FOR SCIENCE (Chapter VIII) have stimulated research on the living sonar of the sea—the dolphin.

We have thought of the sea as a great sewer with an unlimited capacity to absorb our waste products. By ASSESSING MAN'S IMPACT (Chapter IX) we have learned that that capacity has bounds, the ocean and its life are being killed.

In recent years it has been the development of REMOTE SENSING (Chapter X) that more than anything else has enhanced the study of the ocean. Carried aboard satellites in space or attached to buoys moored at sea, remote sensors are gathering information that might otherwise be impossible to obtain. They are rapidly becoming man's eyes and ears.

Nevertheless, there can be NO SUBSTITUTE FOR MAN (Chapter XI). The human observer is capable of insights that no technology can provide. Man in the sea as a free diver or in a research submarine is the best means we have of engendering new questions as we find answers to old ones.

It is essential to the preservation and advancement of civilization that we continue to do basic research in all sciences as they help us to understand the oceans. The urge to gather knowledge simply FOR THE SAKE OF KNOWLEDGE (Chapter XII) is perhaps our most precious human attribute.

The pool of knowledge will be kept brimming if we will foster in ourselves and nurture in our children that most wonderful gift of all—THE GIFT OF CURIOSITY.

Introduction: Outer and Inner Space

Long before the birth of Christ, astronomers contemplated the sky and began to decipher the mysteries of the universe. Centuries before the birth of Galileo, mathematicians and philosophers measured the radius of the earth and plotted its course around the sun. Even the discontinuous atomic structure of matter was theorized. But the oceans were virtually ignored. The bulk of what we know today about the sea was accumulated in little more than the last 100 years.

I prefer to use the term marine sciences rather than oceanography or oceanology, because the study of the sea is not a single scientific discipline: it is an all-out assault, by every available science and technology, on a new medium—the ocean.

The marine sciences first progressed very slowly, by guesswork. The moving shapes observed through the water along the coast or in tidepools, the coincidence between tides and phases of the moon, the animals encountered at sea or caught in nets and traps, the weeds and rocks or mud brought up by anchors, the whirlpools, the reefs and the currents—all suggested the fantastic rather than the rational. Some of the myths about the sea, born in the dark ages and the middle ages, are alive today and occasionally appear as headlines in our newspapers.

Then came the great navigators. Thanks to them, global maps could be drawn, and the prevailing winds and currents around the planet began to be outlined. With the development of the sextant and chronometer, astronomical navigation became the first link between outer space and the sea—a link extended only in the 1970s with the advent of oceanographic satellites. But for the most part, the ships of explorers like Magellan and Cook were carried by oceans of mystery, which they did practically nothing to investigate in depth.

Serious scientific exploration of the sea began with the famed voyage of the *Challenger* in 1872. The main research tool, from the time of the pioneers until today, is—the cable! A line, whether of hemp, steel, nylon or polypropylene, is reeled and unreeled at each "station," lowering and bringing back thermometers, water and bottom samplers, sediment corers, nets and traps, as well as sensors. The sporadic data accumulated in this way was computerized in "world data centers." The picture of the ocean world that one could conjure up from such pin-point measurements was often compared with what Martians might learn about the earth if they were to lower grabs from a flying saucer and bring back a snail, a half-burned cigar, and a sample of polluted air from the exhaust of a car.

Our abstract, instrumental knowledge of the oceans was greatly improved when cameras were lowered on lines, and even more so when bathyspheres and helmet divers brought down the eyes of the explorer. Echo sounding, submersibles, and aqualung divers surpassed the cables as research tools, revolutionized undersea exploration. But their measurements and observations are still analytical and prove insufficient to give us an understanding of the general laws governing the oceans and of what the sea means for the planet and its life.

At human scale, the ocean is immense, its structure extremely complex. Our microinstruments cannot quantify all that goes on in this huge three-dimensional environment.

The first attempt to move from analytical to synthetic knowledge in one field of marine sciences concerned the dynamics of ocean masses. At each "station" of a network involving large provinces of the ocean, temperature and salinity were measured at many conventional depths, and with this data, a simple computation indicated what level the surface would be if the ocean were entirely made of homogeneous "standard" sea water. Assuming that such ideal water masses would move from high levels to low levels, it was possible to obtain a simplified idea of important displacements of oceanic masses. Today this method is supplemented by large, costly, accurate models of the oceans, studied on revolving benches simulating the influence of the earth's rotation and of the coastline and bottom topography.

The effect toward large-scale synthetic understanding of the sea requires synchronizing and monitoring the measurements of many research ships in a given area: the Intergovernmental Oceanographic Commission (I.O.C. of UNESCO) has organized such endeavors in the Indian Ocean, the Caribbean, and other regions. It is the trend of modern marine sciences. A good example is a series of concentrated international expeditions to study the exchanges of energy between ocean and atmosphere in tropical seas: dozens of ships and aircraft, helped by satellites, make studies of all the factors of evaporation in the areas where hurricanes are formed, hoping to indicate how man can better predict and defuse such disastrous tropical storms.

Daily global measurements are badly needed in biological primary production and the pollution of open-sea water. Methods of measuring from high-flying airplanes have recently been developed. But it is still inconceivable to maintain aircraft surveys in all oceans. Efforts are underway to develop instruments, of the same nature as those already tested in airplanes, that could be used by satellites or skylabs of tomorrow. Such satellites would also gather the conventional deep-water data collected by thousands of instrumental buoys anchored in deep water all over the oceans. In the near future, satellites will carry the bulk of oceanographic study; outer space technology is essential to our comprehension of inner space.

Jacques-Yves Cousteau

Chapter I. An Infant Science

Most peoples of the ancient world left little record of what they saw or knew, and we can only guess at their awareness of geography and of the seas around them. Only from the Greeks and the Egyptians, and only from about 1000 B.C., do we have indications of the origins of the marine sciences. We must assume that these records are very incomplete and may be misleading.

The earliest are those of poets—among them Hesiod and Homer. Hesiod referred to land out in the oceans—the Hesperides, the Isles of the Blessed, and others. Homer told of the great voyage of Ulysses around half

---

"Most people of the ancient
world left no record
of what they saw or knew."

---

the Mediterranean. The world was thought to be a relatively flat disc of land around a part of that sea, with a few islands scattered here and there. In the fourth century B.C. Plato gave us a secondhand story of Atlantis.

Undoubtedly the Atlantic Ocean was known to the ancient Greeks. Many early mariners had voyaged from the Mediterranean to reach England, Ireland, and probably Poland as well as the Canary Islands. The Phoenicians are believed to have navigated to the Sargasso Sea, an area in the Atlantic southeast of Bermuda. In the fifth century B.C. Herodotus, the Greek historian, wrote that they had even sailed through the Red Sea, around Africa, through the Strait of Gibraltar, and across the Mediterranean to home. Herodotus also summarized the existing view of the earth as a sphere that was divided into five zones. It was believed that the torrid zone and the frigid zones had climates too extreme for habitation.

In the fourth century B.C. Pytheas, the Greek geographer, made a voyage to the British Isles, determined latitude and longitude, and explained the relation of the moon and the tides. And in the third century B.C. Erastosthenes was able to give the circumference of the earth at approximately what we know it to be—about 25,000 miles—and he suggested that the earth might be circumnavigated if it were not for the vastness of the Atlantic Sea.

At the beginning of the Christian era, the geographer Strabo is said to have measured the Mediterranean to a depth of more than a mile at one point, but no one knows how he did it. In the second century A.D. Ptolemy made a map of the world that gave the circumference of the earth at 18,000 miles. His map was the standard for a very long time, and it did much to encourage the idea of reaching the Indies by a westward voyage. In Rome the naturalist Pliny studied the sea and its life, mostly at secondhand, and compiled his great work *Historia naturalis*.

Then darkness fell. For a thousand years classical learning was ignored, and the geography of the world was interpreted by reference to scripture. The earth was flat again. Yet through these long centuries of intellectual blindness, there continued to be daring voyages on the Atlantic, most notably by the Vikings, who traveled to the Faroes, Iceland, Greenland, and America. Their discoveries, however, remained largely unknown to the rest of the world.

*One of the most popular **misconceptions of the sea** was that it was the domain of all sorts of hideous monsters. The notion must have struck a terrible fear in the hearts of ancient mariners and may have been the cause of many a wreck.*

*Dutch mariners study the principles of **scientific navigation** in this engraving for the title page of an atlas of the sea published in the Netherlands in the seventeenth century.*

Laying the Groundwork

By the end of the seventeenth century, the groundwork had begun to be laid for a truly scientific study of the sea. The English physicist and philosopher Sir Isaac Newton had advanced an explanation of the tides and his work had been elaborated upon by the French mathematician Pierre Simon de Laplace. In 1776 the French chemist Antoine Laurent Lavoisier published an analysis of seawater. The British scientist Charles Blagden explained the relation of the variable freezing point of water to the concentration of dissolved substances it might contain. The English chemist Robert Boyle had

taken an interest in the salinity of the sea, and even in the late seventeenth century he realized that the salinity in any part of the ocean depends on the amount of evaporation and rainfall—the more evaporation, the greater the salinity. In England, too, Robert Hooke, who discovered that living creatures were made of cells, delivered a number of lectures on methods that might be used for deep-sea research.

These early scientists generally concerned themselves with one characteristic of the sea, and it was most often an aspect, such as salinity, that could be explored in the laboratory. But these scientists laid a necessary

*Early scientists of the sea were usually concerned with a single characteristic that could be studied in the laboratory, but they laid the foundation for those who would later study **the sea** firsthand.*

foundation for those who would later seek to understand the interdependences of these discoveries and to make new ones through close observation of the sea itself.

In 1802 the first theory of waves in deep water was published. In 1812 Alexander von Humboldt, the German scientist, explained that the very cold water at the bottom of the tropical oceans was evidence of bottom currents flowing from the polar regions toward the equator. He further made a distinction between the general circulation of the water in the ocean and the rapid surface currents that had most bearing on navigation. In 1820 Alexander Marcet, an Englishman, observed that water from different parts of the ocean contained the same ingredients and that they existed in nearly the same proportions. In the same year William Scoresby, an arctic explorer and scientist, published the results of his intensive investigations of the behavior of the whale. In 1832 the first study of currents in the Atlantic Ocean was published.

Early oceanography was also acquiring a certain amount of popular appeal as natural historians were more and more drawn to the sea and described their discoveries as did William Spiers in his tremendously popular *Rambles of a Naturalist.*

Milestones in Making the Map of the World

From earliest times every highly cultured people has been eager to chart the earth. Marine charts were military secrets. Over thousands of years countless discoveries of seas and rivers, islands and continents, deserts and forests have brought us very close to that goal. Now our space satellites are helping us to fill in the last missing pieces.

Along the way a few great documents have marked the stages in the pursuit of that goal. Perhaps the first of them was the map of Ptolemy, an astronomer and geographer, a citizen of Alexandria, who lived from about 87 to 150 A.D. He knew a great deal about the Mediterranean, which is not surprising; the greatest explorers of ancient times had been the Cretans, the Greeks, and the Phoenicians, all sailors on that sea. What is surprising is that Ptolemy also knew of the

A 16th century sailing vessel. The navigator on the quarterdeck is relating the inclination of the sun's rays to the reading of the magnetic compass.

British Isles and Ireland. The knowledge had come to him from Pytheas of Marseilles, a daring Greek merchant who had sailed all the way to the fringe of the arctic. But the New World was unknown, Africa was almost a mystery, and the Indian Ocean was thought to be an inland sea.

During the medieval period geographical knowledge in Europe dwindled along with most scientific learning. However, some important discoveries were being made by the Vikings, who, in the ninth and tenth centuries, sailed eastward to the White Sea and westward to Iceland, Greenland, and America. Their influence on geography was small, however, because those discoveries were not brought to the attention of the rest of Europe. More significant for the future was the revival of interest in the ideas of the Greeks.

At the time of the intellectual and artistic reawakening in Europe that we call the Renaissance, there was a revival of interest in exploration and in the development of means to carry out far-ranging voyages by sea. The invention of new instruments of navigation and the improvement of an old one—the compass—as well as great strides in the art of shipbuilding made this possible.

The first of the notable voyages was the rounding of the Cape of Good Hope and the return to Portugal by Bartholomeu Diaz in 1487. Five years later Christopher Columbus's three caravels arrived at Guanahani, and America had been rediscovered. In 1499 Vasco da Gama sailed around the Cape of Good Hope and all the way to India. And Vasco Nuñez de Balboa discovered the eastern shore of the Pacific in 1513. In 1521 Ferdinand Magellan led the first Europeans across the Pacific, and in 1522 one of the vessels of his fleet completed the first circumnavigation of the globe when it returned to Spain. Magellan himself had been killed in the Philippines, but his has been called "the most astonishing voyage that ever was made." It is further notable because Magellan is believed to have made the first soundings ever taken in the deep seas. During the voyage he lowered handlines to a depth of perhaps 1200 feet. After Magellan, no deep-sea soundings were taken for about 300 years. A succession of explorers followed Magellan into the Pacific, and the greatest of them was James Cook, who discovered the Hawaiian Islands and many others. He circumnavigated Antarctica without seeing it and tentatively assumed that the "continent" called Terra Australis was no more than a myth.

In 1568 the first map using what would come to be called the Mercator projection appeared. It would be used more than any other method of rendering a map in two dimensions. Gerardus Mercator's *Atlas* began to appear in 1585. His "magnetic mountain," rumored to draw nails from ships' timbers, can be forgiven. The *Atlas* was a remarkable achievement for its time.

The arctic had made its appearance quite early in European history as a dim, legendary region. Now it lured Martin Frobisher through its waters in 1576–78. John Davis followed him in 1585–87, and Henry Hudson and William Baffin in the early part of the seventeenth century. More than a hundred years later, Vitus Bering and Frederick Cook crossed the same frozen paths searching for a Northwest Passage to the Pacific.

For most of these men the sea served as a highway. Some of them, like Magellan and Cook—"that truly scientific navigator"—turned their attention to questions of the sea itself as they made their voyages of geographical discovery. All of them helped to bring us finally to new frontiers—the exploration of sea and space.

A chart of the Gulf Stream, published in 1770 under the direction of Benjamin Franklin.

The Secret of the Gulf Stream

A man of insatiable curiosity, Benjamin Franklin had studied the "river in the Atlantic" on crossings from colonial America to England. He had discussed it with his relative, Captain Folger of Nantucket Island. Captain Folger and the Nantucket whalers intimately knew the powerful stream that came up from the south and then veered off to the east. They had followed whales along the edges of it and noted that a ship that got into the current was carried off very rapidly. They had met and talked with masters of ships from England who were trying to stem the stream, fearing that to go north to avoid it would put them on Georges Bank or the Cape Sable or Nantucket shoals. Their voyages were delayed even in good winds because they lost 60 or 70 miles of way to the currents each day.

When he was appointed postmaster general of America, Franklin asked Folger to mark the stream on a chart. In 1769 he sent it and Folger's directions for avoiding both the stream and the dangerous banks and shoals to his superior, the Secretary of His Majesty's Post Office. Franklin suggested that the chart and directions be distributed to the packet captains to shorten their voyages.

The Honorable Secretary agreed, and the chart was published in 1770. It was soon in use by all Atlantic navigators. Ships that cross from Europe to America today still need to know the meanderings of the Gulf Stream, but now the charting is done from day to day by space satellites.

Sailing by Stick Chart

Most islanders of the South Pacific have long been highly skilled navigators. They sailed their canoes across vast stretches of ocean, often out of sight of land for several days at a time, but they always knew where they were going.

There were a number of methods they used to chart their course—the apparent movement of the sun and stars, direction of the wind, known ocean currents, and the flight of migratory birds. Some of them made an effective use of another method—wave patterns on the surface of the sea made by reflection against and refraction around islands and atolls. The Marshall Islanders and other Micronesians were particularly skillful at this navigational device, and young men were instructed in the method with the aid of stick charts.

These charts, made of reeds and shells and often very beautiful, mark the locations of atolls and islands and show the movement of waves as they are bent or altered by bodies of land. With an understanding of the interaction of several prevailing swells and their refractions, it could be determined in which direction land lay. An understanding of the use of the stick charts was a closely guarded secret for a long time, known only to the master navigators and the initiates. An interesting book about the native Pacific sailors, *We, the Navigators*, was written by the solitary navigator Dr. David Lewis.

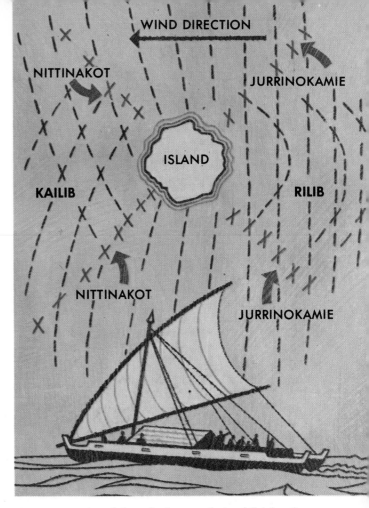

The illustration (above) shows what might be depicted on mattang-type *or instructional* **stick charts** *common to Micronesia (below). In this case, it shows the influence an island would have on ocean swells as one approached the island from various direction. A* rilib *situation indicates reflected waves and shows that land lies in the direction from which the waves come. As the navigator travels to the side of the rilib, he eventually finds the reflected waves are no longer parallel to the main swell. This area, the jurrinokamie, is characterized by a choppy interference pattern which can be followed to the island. The refraction of waves causes the kailib, which tells the navigator that land lies in the direction from which the waves are coming. In the regions of the nittinakot, refracted waves have no influence on the prevailing swell, allowing the navigator to follow the line to land.*

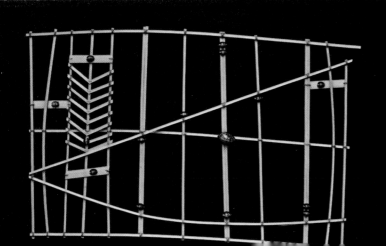

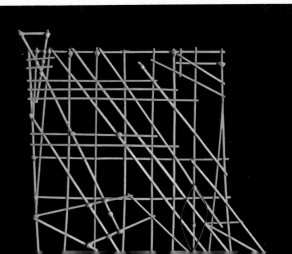

The Pathfinder

While sailing on the *Falmouth* around Cape Horn in 1831, Matthew Fontaine Maury, a young lieutenant in the U.S. Navy, wondered why there were no charts showing the direction and speed of the winds and currents. Couldn't the sailing time be shortened considerably if they were known and used to help the ship rather than hinder it?

In 1842 Maury was put in charge of the Depot of Charts and Instruments, which later became the U.S. Naval Observatory and the Hydrographic Office. He had set himself the goal of "nothing less than to blaze a way through the winds of the sea by which the navigator may find the best paths in all seasons." In his new position he was able to realize that goal. He asked ships' captains all over the world to send him their observations of winds and currents that were encountered day by day at a particular latitude and longitude—a description of the same route in all seasons. They also sent him their measurements of temperatures, barometric pressures before, during, and after a storm, and news of other phenomena such as fogs, whales, birds, islands, and shoals sighted, or of errors in earlier charts. With this information and with the help of old logs he compiled the first systematic collection of such data—*Wind and Currents Charts*. These charts were sent all over the world, and they helped to shorten voyages by as much as one-third and to make them safer as well. The average sailing time between Rio de Janeiro and New York was cut from 55 to 35 days by using the newly available knowledge of currents and winds. Between New York and San Francisco around Cape Horn, a distance of 14,000 miles, the time was cut from 183 days to 135 days. Nine years after the publication of Maury's first chart, a thousand navigators were at work

collecting weather and current data according to a uniform plan. New charts that were published represented more than a quarter of a million days of observations.

It was only natural that Maury should have asked what caused the winds and the currents that had now been charted, and he answered his questions in his more famous work, *The Physical Geography of the Sea*. This large text was the first classical work of modern oceanography.

The book described the Gulf Stream and how it modified climates and affected the weather and determined the number and

Few sailing ships still transverse the oceans of the world. Modern, efficient navigation requires weather and current maps, called "pilot charts."

quality of fish in it. He concluded that the forces that moved the Gulf Stream were the winds, the earth's rotation, and the differences in the specific gravity or relative density of the waters. He was incorrect in the last point, and he did not know—nor could he have known—that topography was also a conditioning factor.

More importantly, the book presented a theory of the salinity of the sea that contributed to an understanding of the sea as having a complex circulatory system which mixed the waters of seven seas. It was from his understanding of the general uniformity of proportions of dissolved matter that

Maury was able to sense the sea as a great dynamic system of integrated and worldwide circulation.

Maury's findings provided the nucleus from which our present understanding of the dynamic processes of the ocean has grown. Study of the physical interrelationships of air and water which govern oceanic circulation has been the primary concern of physical oceanography ever since.

Chapter II. Pioneers in Ocean Sciences

With Captain Cook's last voyage in 1776–79, the general outlines of the oceans had been revealed, but relatively little was known of the oceans beneath the surface. Almost nothing was known of the sea floor.

The first real success at sounding the ocean depths was achieved by Sir John Ross in 1818 when he obtained a measurement and a bottom sample at 6300 feet in Baffin Bay off the west coast of Greenland. Soundings were advanced in 1854 by a device that had a detachable weight that dropped off when the bottom was struck. The falling off of the weight made it easier to know the bottom had been reached because after a few thou-

---

"It is necessary to look forward to a harvest, however distant that may be, when some fruit will be reaped, some good effected."
Charles Darwin

---

sand feet of line had been paid out, the line's weight was so great that a man could not tell when the end had touched the bottom. Another important advancement came in 1870 with the introduction of steel cables.

In 1855 Matthew Maury published his *Physical Geography of the Sea,* one of the first books about oceanography written in English. William Ferrel, an American who was intrigued by Maury's book, became the first man to explain the motion of the surface waters as being due to winds.

Charles Darwin's voyage on the *Beagle* from 1831 to 1836, though not strictly oceanographic, did much to stimulate interest in the study of the ocean.

Deep-sea research is generally considered to have begun with the voyage of the *Chal-*

lenger in 1872–76. Under the command of Sir C. Wyville Thomson, the *Challenger* sailed around the world collecting information in many areas of ocean sciences. She made 492 deep soundings as well as 133 dredgings and made other measurements at no fewer than 362 stations. At these stations, data was collected on weather, currents, water temperatures, water composition, marine organisms, and bottom sediments. This voyage was one of the great achievements of scientific investigation, and it also served to reveal how little we knew about the sea.

The *Challenger* Expedition inspired many nations to similar feats. The German ship *Gazelle* circumnavigated the world in 1874–76 and so did the Russian research vessel *Vityaz* in 1886–89. The Austrian ship *Pola* made observations in the Red Sea and the Mediterranean in 1890–98; the American vessel *Blake* worked in the Caribbean in 1877–80 under the direction of Alexander Agassiz; and the Norwegian ship *Fram* studied the arctic region under Fridtjof Nansen.

But a good overall picture of the oceans was not obtained until the early twentieth century. The general topography of the ocean floor was gradually mapped. The first detailed studies of a particular part of the ocean was made by the *Meteor* Expedition in 1925–27. The *Meteor* made 14 crossings of the South Atlantic collecting data day and night in all weather and seasons. It was the first to use an electronic echosounder to measure the depths, and it made more than 70,000 such measurements. These soundings made clear how rugged the ocean floor is.

A plankton net is lowered over the side of the Hirondelle, the research ship of Prince Albert the First of Monaco, one of the great pioneers in oceanographic research in the nineteenth century.

The Challenger

In 1871, Wyville Thomson, a Scottish naturalist, was asked to make a trip around the world as director of an expedition to study the deep seas. An old sailing vessel was bought for the expedition. It had an appropriate name—the *Challenger*. Thomson was ordered to find out "everything about the sea." This was to include both the biological and physical conditions of the oceans. The expedition was to record temperatures both at the surface and in deep water, study the movements of the tides and the currents, chart the relation of barometric pressure to latitude, record the chemical content of the water, and gather and classify specimens of animals and plants at all levels of the ocean.

The expedition set a model for almost every oceanographic voyage from that day to modern times.

The expedition dredged the bottom everywhere, and everywhere they brought up ooze—the sediment made of the skeletons of tiny animals that once lived in the surface waters. They collected 4000 species of plankton, recognizing that these mere specks of life are a very important source of food for the larger animals of the sea. And 3508 new species of radiolaria were collected to add to the 600 that had been known until then. In all, they discovered 715 new genera and 4417 species of living things, demonstrating that the ocean was teeming with life of all kinds. It proved, as well, that life existed at

The modern science of oceanography was born when **H.M.S.** **Challenger** *made its historic expedition, which covered 68,890 miles.*

In the natural history workroom aboard the *H.M.S.* Challenger *(above), biological specimens could be examined and studied in detail.*

Dredging and sounding arrangements on the Challenger *(right). The expedition was to learn "everything about the sea."*

The scientists of the *Challenger* discovered that it was in fact a precipitate that was formed when the specimens were placed in preservative. It had been one of experimental science's most notorious blunders.

In its 68,890 miles Challenger established the main contour lines of the ocean basins and the first systematic plot of currents and temperatures in the sea, and demonstrated that the temperature of deep water in each zone was fairly constant in all seasons. The findings of the expedition were so extensive that it took 50 volumes to describe them. The science of oceanography had been born.

great depths. It was determined that some animals are not affected by the water pressure at different depths because their tissues are permeated with liquid about as incompressible as the water.

The expedition made a standard series of observations everywhere the ship stopped —total depth of water, temperatures at various depths, atmospheric and meteorological conditions, and direction and rate of current on the ocean surface and occasionally of the currents at various depths. During the three-and-a-half-year voyage, they made observations at 362 stations.

One of the expedition's most important accomplishments was the shattering of a theory of T. H. Huxley and Ernst Haeckel. They had claimed that at least a major part of the ocean floor was covered with a thin layer of almost structureless living slime that Haeckel called "Bathybius"—the primordial protoplasm. It was considered to be the root of the evolutionary tree, the simplest form of life from which all others had evolved.

The Princess Alice was one of the very first ships to conduct oceanographic research in the 1890s.

great saving in stowing space on board and a considerable saving in the time needed for lowering and hauling in the nets. The *Challenger* had taken more than two and a half hours to make a sounding at 2435 fathoms. The *Blake* took about an hour for a sounding at 2929 fathoms. Another innovation was the double-edged dredge that could land on either side and still work.

The *Blake* made three cruises under Agassiz's direction, from 1877-80. Agassiz considered their discoveries to be "but little inferior to those of the *Challenger*." One of the most exciting discoveries that had been made was that the dredged materials suggested that at one time the Caribbean and the Pacific Ocean had been connected.

The U.S. Fish Commission later asked Agassiz to command a series of investigations to be made by the vessel *Albatross* on the Pacific side of Panama, which resulted, among other things, in some extraordinary collections of marine life.

Agassiz's greatest contribution was probably in his study of coral reefs. He spent 30 years observing them. While he was studying coral reefs in the tropics, another great pioneer,

*Bulging eyes of **rockfish** (below) are caused by expanding gasses as the fish ascends from the deep.*

Agassiz and Nansen

In 1877 the director of the U.S. Coast Survey asked the young Swiss-born Alexander Agassiz to take charge of a series of dredging cruises on the schooner *Blake*. The *Blake* surveyed the Gulf of Mexico, the Caribbean, and the waters off the east coast of the southern United States. Agassiz introduced some important changes in dredging methods. Instead of the ropes that the *Challenger* had used, he used steel cables, making a

24

Fridtjof Nansen, was icebound in the frozen arctic aboard the *Fram*. The Norwegian polar expedition was undertaken from 1893 to 1896. One of Nansen's primary objectives was "to form a more complete idea of the circulation of the northern seas." To do this, the *Fram* was constructed to withstand the crushing pressures of readjusting ice. The bow and stem of the *Fram* were built of oak four feet thick, which successfully protected it from being crushed. The ship's log read, "It looks as if we were being shut in . . . the ice is pressing and packing around us with a noise like thunder. It is piling itself up into long walls, and heaps high enough to reach a good way up the *Fram*'s rigging; in fact, it is trying its very utmost to grind the *Fram* into powder."

During the endless days of mandatory inactivity while trapped in the creeping arctic ice pack, the crew methodically took their measurements. One of their most startling discoveries was that the arctic sea is not shallow but in excess of 2000 fathoms. In addition to depth measurements, they made meteorological observations, recorded water temperatures and salinity, and noted the variety of benthic animals and even plankton with which Nansen became fascinated and devoted much of his time.

Nansen grew restless in his sendentary existence. He felt that their inactivity was neither life nor death but a state between the two. After considerable deliberation, he decided to set out on foot for the North Pole. Nansen and his companion met with extreme difficulty as they trudged north. Their clothes became soggy and heavy with moisture, sleeping bags became unbearably heavy, lifting sledges over hills of snow became impossible, fatigue prevented even the simple act of eating. Food was getting scarce, signs of life were nonexistent, and some of the dogs had died by the time Nansen decided they must turn back. With all sense of purpose gone, Nansen wrote after killing a seal, "Here we lie far up in the North, two grim, black-soot-stained barbarians, stirring a mess of soup in a kettle surrounded by ice and nothing else." They eventually reached land and came upon another arctic expedition and returned to Norway. Within a few days Nansen received word that after 35 months the *Fram* had made it free of the imprisoning ice by using dynamite to hack out an escape route.

A small octopus is removed from an assemblage of animal life dredged up from the depths of the ocean.

Small crustaceans, as well as red rockfish and flatfish, make up most of the catch.

A Royal Patron

A man who devoted the greater part of his fortune and energies to the study of the oceans and who furthered and encouraged the work of others was a prince. He was Prince Albert I of Monaco, and his work stirred interest in all areas of oceanography.

In 1885 Albert acquired a ship called the *Hirondelle* and began a series of cruises in the Mediterranean and the North Atlantic that were to make a great contribution to ocean studies. He was early concerned with the question of life at the middle depths, and to this problem he brought new and unprecedented equipment and techniques. The prince and the scientists who advised him believed that fish, squid, and other fast swimmers inhabited all levels of the sea but were not caught because the trawls used were too slow. The prince designed a high-speed trawl and the hypothesis was proven true. Among the unique techniques he employed was the examination of the regurgitation of dead whales.

On the *Hirondelle*, Prince Albert made studies of the Gulf Stream. The cause of ocean currents was still being debated, and the Gulf Stream was the example that proponents of different theories drew on. One of the questions that was much debated was the extent of the Gulf Stream. Some insisted that the current never crossed the Atlantic, while others maintained that it did, dividing somewhere in the North Atlantic, sending one arm north toward the British Isles and another south toward the Iberian Peninsula. There had been attempts to answer the question experimentally, but by 1885 no conclusive answer had been given. Prince Albert determined to finally settle the question. The *Hirondelle* sailed into the Atlantic and a few miles northwest of the Azores, released a cargo of 169 floats, each of which carried a message in 10 languages. The following year 510 bottles were set adrift, and on the last of the drift-bottle cruises 931 floats were released between the Azores and th Grand Banks of Newfoundland. Of the 1675 bottles set afloat, 227 were recovered, and their position seemed to indicate the existence of both a clockwise gyre in the North Atlantic and of a current (the North Atlantic Drift) which branched northward from it. The prince's experiment had given considerable weight to those who had argued for a transatlantic Gulf Stream.

Prince Albert's energies were by no means confined to the study of currents. Within the next 25 years he had devoted his attention to biological oceanography, bathymetry, marine meteorology, and education. He built and worked from three ships more modern than his first: *Princesse Alice I, Princesse Alice II,* and *Hirondelle II.* In 1910 the Oceanographic Museum which he had planned and built was dedicated at Monaco. It housed aquariums stocked with fish from many seas, collections of preserved animals, and exhibits of oceanographic equipment. There were also laboratories, a library, conference rooms, and a research ship.

The Princess Alice II, *Prince Albert I's third oceanographic research ship, approaches a school of pilot whales in the calm waters around the Azores.*

Hirondelle II, *one of several oceanographic research ships of* **Prince Albert the First of Monaco,** *a pioneer in marine science, lies at anchor in the harbor of the principality in 1911 (above). The Prince (below) looks on as other scientists examine a catch of biological specimens and make sketches, some of which are seen at right. Prince Albert interested himself in all aspects of ocean studies and furthered and encouraged the work of others.*

In 1911 Prince Albert established the Oceanographic Institute in Paris, which had professorships in marine biology, physiology of marine life, and physical oceanography. Students were to receive formal training at the institute and field experience using the facilities of the museum at Monaco. Both institutions published scientific journals. They were the first periodicals devoted exclusively to oceanographic studies. The prince himself published a long series of volumes describing the collections he had made.

Into the Realm of the Pharoahs

Corals (above), *genus* Tubastrea, *help construct the most remarkable structures of the sea.*

The Great Barrier Reef of Australia, the largest of coral reef formations and one of the most remarkable natural constructions in the world, has long intrigued scientists. The extraordinary formation, which extends for 1250 miles off the northeast coast of Australia was probably first sighted by the Spaniard Luis Vaez de Torres, who commanded one of the ships composing a squadron that was sent out from South America in 1605 to explore the South Pacific. His discovery was kept a close secret, however, until in 1762 the British captured Manila and found a full account of the discovery in the Spanish archives. Captain Cook, however, was the first to begin exploring the reef. After charting the east coast of Australia for the first time, he worked his way northward between the reef and the coast until his ship, the *Endeavor*, ran aground. When the ship had been made seaworthy, they set sail and continued on to Torres Strait. A more detailed exploration and charting of the Great Barrier Reef was done by William Bligh—famous also as commander of the ship *Bounty*—and Matthew Flanders. The most important early work on corals themselves was done by Charles Darwin, which he published in *The Structure and Distribution of Coral Reefs*. Our knowledge of corals was further greatly advanced by F. Wood Jones, who spent 15 months on the coral island of Cocos-Keeling in the Indian Ocean and then published his *Coral and Atolls*. But the first time that science really came to grips with the coral reef ecosystem was in the Great Barrier Reef Expedition of 1928-29.

*A section of the **Great Barrier Reef of Australia** (above), the largest coral reef in the world.*

Yonge himself wrote a long and popular description of the expedition and its findings called *A Year on the Great Barrier Reef* in which he said, "Corals are the greatest of the world's builders, and the Great Barrier Reef is the most magnificent of their creations." It seems strange now that at the time Yonge had to dispose of a popular notion that the coral was an insect. "The coral 'insect' is as mythical as the griffon," he wrote.

The Great Barrier Reef Expedition not only put the study of the coral reef ecosystem on a firm basis but also marked a scientific watershed between an age when expeditions made great collections with the object of simply describing them and the modern age of experimentation and explanation.

*The Luana (below) was the platform for scientists studying the Great Barrier Reef in 1928-29. **Tenders** (bottom picture) pump air to a diver and keep his lines untangled.*

This British undertaking comprised a biological and a geographical section. C. M. Yonge was in charge of the biological group. He was particularly concerned with the feeding and digestion of corals and other reef animals and with the significance of their symbiotic algae in the life of corals. A member of the biological team worked primarily on "borers" into coral. Another studied the manner in which corals form their skeletons. Yet another scientist devoted himself to the study of the phytoplankton and the effect of sediments on corals. The different aspects of the ecosystem were also studied until an overall view was put together.

The report of the expedition was published in four volumes by the British Museum.

Beebe

The development of ocean sciences in the United States was greatly stimulated between the two world wars by the work of William Beebe. According to one writer, Beebe "was responsible, through his explorations, writings, and lectures, for creating public awareness of the existence of a field of scientific endeavor called oceanography." Beebe's work generated much interest in the oceans at home and abroad, especially as submarine warfare revealed that much more knowledge of the depths was required. Beebe's celebrated bathysphere, probably more of an engineering feat than a scientific triumph, focused attention on exploration of the deep sea and encouraged the development of nonmilitary minisubmarines dedicated to underwater research.

The bathysphere made its first descent in 1930 with Beebe and Otis Barton, who had designed the chamber, aboard. The bathysphere was a steel ball, four feet nine inches in diameter with walls one and a quarter inches thick and weighing 5400 pounds. It had three windows made of fused quartz, the strongest material then known that was capable of transmitting light of all wavelengths. Through one window a searchlight was directed into the water. The bathysphere was supported and serviced by a ship at the surface.

Half-Mile Down is the record of the work of the bathysphere and its two explorers. One thing that interested Beebe about the descents were the lights and the colors. In a chapter called "A Descent into Perpetual Night," he describes the first time the bathysphere touched bottom: "At 11:12 A.M. we

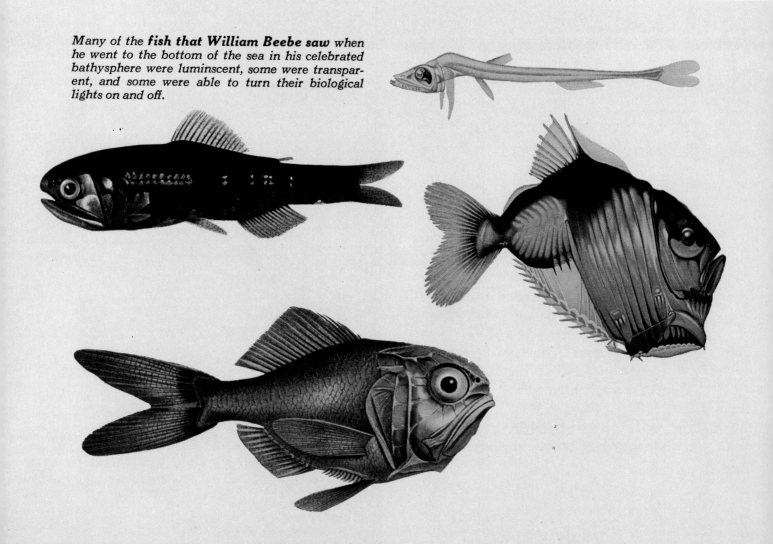

*Many of the **fish that William Beebe saw** when he went to the bottom of the sea in his celebrated bathysphere were luminscent, some were transparent, and some were able to turn their biological lights on and off.*

came to rest gently at 3000 feet, and I knew that this was my ultimate floor, the cable on the winch was very near its end. A few days ago the water had appeared blacker at 2500 feet than could be imagined, yet now to this same imagination it seemed to show as blacker than black. It seemed as if all future nights in the upper world must be considered only relative degrees of twilight. I could never again use the world BLACK with any conviction."

But that black world was not without its light. At 2000 feet there were consistently a number of bioluminescent organisms within Beebe's view. A number of previously unidentified fish swam by with large cheek lights. There were hatchetfish, anglerfish, jellyfish, eels, and even snaillike animals, with fleshy extensions called pteropods, used to fly through the water——and they all glowed with their own colored lights. He was the first man to witness the deep-sea shrimps which, instead of discharging ink like the octopus to confuse a predator, extrude liquid fire, a luminescent material. Myriads of fish displaying light organs along their sides appeared to Beebe like miniature passenger trains in the night. He was overwhelmed by the living fireworks display to the point of using up all the dramatic adjectives he knew, leaving him, after a time, unable to express adequately what he was seeing. The record of his narrative at times becomes dull and lifeless; he would return to the deck of the ship speechless, as though in a trance somewhere between fancy and reality. He concludes that the only other place comparable to these marvelous regions must surely be naked space itself, between the stars far beyond the atmosphere.

Chapter III. Recent Voyages of Discovery

After the *Meteor* Expedition of 1925–27, oceanographic research became more systematic and thorough. Equipment and techniques rapidly advanced in sophistication. A number of important research laboratories were established. An important compendium of most of the knowledge we had gained of the sea to that time was published in 1942 in the United States by Sverdrup, Johnson, and Fleming. It was called *The Oceans,* and is still a standard reference book.

After World War II technological advances gave a tremendous boost to oceanography. Since the war there has been an increasing trend toward international cooperation in large-scale oceanographic efforts. No single country has monopolized study of the sea, and in recent years even the most important

---

"With the Meteor Expedition, oceanographic research became more systematic and thorough."

---

explorations have not been the private preserve of the larger nations. Furthermore, the great expeditions have not concentrated on any one aspect of the sea, but have usually concerned themselves with the acquisition of data in all areas of oceanography.

Recently, too, many of these expensive expeditions have been so spectacular that they have stirred the interest of the communications media which, in turn, have begun to inform the people at large of our new knowledge of the oceans. The first project to receive this kind of attention was probably MOHOLE, the attempt to drill through the crust into the earth's mantle. This seemed possible because the crust of the earth is much thinner under the sea than under the continents. The ambitious goal was never achieved, but the techniques which were then developed provided the foundation for a very successful deep-sea geological drilling venture. JOIDES, the Joint Oceanographic Institutions' Deep Earth Sampling Program, has the practical objective of drilling and sampling the upper layers of the earth's crust. Its initial effort in 1965 off the coast of Florida was such a success that it inspired the Deep-Sea Drilling Project, an effort to drill many holes in the Indian, Pacific, and Atlantic oceans. Begun in 1968, this project has also met with considerable success.

Another recent advance in marine technology has been the design and construction of ships especially for scientific work at sea. One of the first such vessels was the *Atlantis,* built by the United States in 1931. Prior to this, research ships had been converted from ships designed for other purposes.

One of the great spurs to study of the seas was a national disaster. The loss of the nuclear submarine *Thresher* in 1963 demonstrated that the sea remains a dangerous adversary that demands better understanding.

The prospect of man's living in the sea has always intrigued scientists and we have made considerable advances toward this end, putting man on the continental shelves for weeks at a time, most notably in our own Conshelf projects and in the Sealab projects of the U. S. Navy. Nevertheless, it remains ironic that although we have set foot on the moon, we have yet to walk on the bottom of the ocean's depths.

*A small **midwater trawl,** one of the biological oceanographer's most valuable tools, is raised from the sea with its sampling of life beneath the surface.*

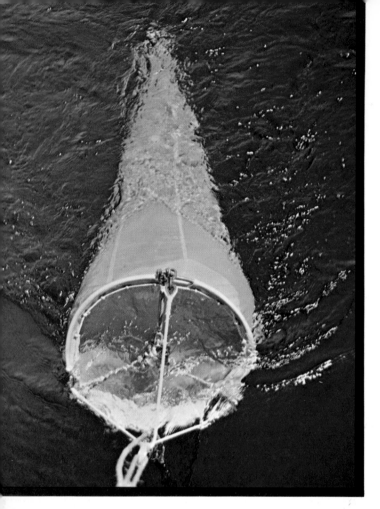

kind of plankton that is the staple of most whales. He constructed a device called a continuous plankton recorder that could be towed behind a boat at full speed at any required depth. It could sample the plankton mile after mile and thus give a continuous record of the changes in its composition. Never before had such an extensive plankton survey been conducted.

Discovery II, built in 1929, continued the investigation of antarctic waters. The voyages are described in *The Open Sea*.

*Microscopic copepods (above and below) can be collected in the fine-mesh **plankton net** seen at left.*

In Antarctic Waters

For centuries man has been exploring the arctic, but interest in Antarctica and its waters had long lagged behind. In 1925, however, Professor Alister Hardy and other British scientists sailed on the *Discovery* for the South Atlantic to assess the relation between whales and plankton. They studied the physiology of the whale itself and learned a great deal about the reproductive patterns and feeding behavior of these mighty creatures. They studied the whale's environment—the conditions that determined the whale's way of life—especially migrations and the effects of chemical, physical, and biological factors. They discovered how whales sleep—usually below the surface with their blowholes tightly closed, holding their breath.

Hardy undertook to investigate the conditions underlying the production of krill, the

34

Abyssal sea cucumbers and brittle stars indicate a rich rain of material is forming these sediments.

The Albatross

On July 4, 1947, the Swedish Deep Sea Expedition set out to turn back the pages of geological history. Their ship was the *Albatross,* and the chief scientist aboard was Professor Hans Pettersson. The expedition made a survey of the bottom of the sea at all depths and studied the deposits found, their interaction with the ocean waters, and the thickness of the sediment carpet. The expedition sampled and analyzed water layers from surface to bottom, noting the temperature and salinity, the amount of dissolved oxygen in the water, the nutrient salts, and the general biological conditions. The expedition also searched for radioactive elements —radium and uranium—and, most importantly, took many sediment cores. For this last purpose it was equipped with a new piston core sampler—a long steel cylinder with a piston inside, operated by water pressure. A winch lifted it from its horizontal position on the deck. Then it was lowered through the water and into the sediment. The piston inside the coring tube remained in contact with the sediment while the heavy core permeated the bottom, forcing a column of sediment upward into the cylinder.

It was estimated that the Atlantic sediment often extended 12,000 feet below the floor itself and that it would have taken more than 500 million years for this much sediment to settle. These figures were later disproven with better techniques and studies on the spreading of the sea floor.

The *Albatross* took 300 samples from the Atlantic, Pacific, and Indian oceans and the Mediterranean, some of them reaching sediment layers deposited several million years ago. During the entire cruise, the *Albatross*'s echo sounder made continuous recordings of the contours of the sea bottom, showing in many places a rugged topography. Only recently have observations proved that no echo-sounding technique, even the most advanced, could give an idea of the jagged relief of mountainous provinces that have never been softened by atmospheric erosions.

The expedition brought to light far more than had been known about the age of the sediments. It also provided a demonstration of the relationships among sciences.

Some bottom dwellers are sessile (above). Some *spew up substrate as they burrow (below).*

In the Canyons of the Sea

Someday, somehow, man will be able to visit the canyons of the sea with greater ease than he can the Grand Canyon in Arizona, because the neutral buoyancy of his crafts will endow him with three-dimensional maneuverability. Only the limited range of vision under the sea will deprive visitors of overall contemplation possible on land sites. However, undersea canyons exist, and are often deeper and steeper than those in our dry world. As of today, an accurate picture of the undersea ranges has to be figured out in our imaginations, from inaccurate maps.

Valleys of the sea floor, such as the Hudson Canyon, were first described a little over a century ago. They were mapped with the aid of soundings taken from the surface, and immediately speculation about their origin arose. They were first explained as submerged river valleys. Some suggested that ocean floor currents had produced them. Around the turn of the century it was proposed that they were formed as a result of the great continental uplifts that brought about the Pleistocene glaciation.

A series of echosound profiles of the eastern continental slope of the United States was made by the Coast and Geodetic Survey around 1928 which revealed that there is a whole series of canyons there and that they extend to depths of a mile and more.

This mapping revived the interests of scientists in canyons, and new hypotheses about their origin sprang up. One said that they were caused by the outflowing of bottom currents initiated by the buildup of waves along the coasts. Another claimed that turbidity currents were responsible for the formation of the majority of them. Still another said they were caused by artesian sapping; and another suggested that they were a byproduct of tsunamis. All the scientists who put forth their hypotheses had one thing in common—they had never taken a firsthand look at the canyons themselves. The first known dive into a canyon for scientific reasons was not made until 1947 and, of course, it was in hard-hat equipment. Two years later, however, free diving began, and with it canyon studies took a new departure. Then, in the 1950s, our diving saucer, the first highly mobile, compact submersible, went down into a canyon and suddenly the

Diving Saucer (above) is a two-man submersible equipped with a mechanical arm, viewing ports, strong lights, and a camera system. It carries the crew to 1000 feet beneath the surface.

Puces (left and right), are maneuverable one-man mini-subs capable of diving to 1600 feet.

studies were greatly accelerated. The range of operations had been greatly expanded. The saucer enabled us to descend deeper and stay longer than the free diver while still being able to move about, look around, take pictures, and even pick things up. Scientists at Scripps Institution of Oceanography have used the saucer extensively down to 1000 feet in the two canyons off La Jolla.

The goals of canyon research are still to determine their origin and to discover the processes that are still active in carving their beds deeper. This knowledge can contribute a better understanding of marine erosion and sedimentation in the geological column. Canyon studies can also be of benefit to the petroleum geologist.

Studies of canyons carried out from submersibles like the diving saucer have included current measurements, tests on the properties of sediments in the fill, and investigation of the structure of sediments from large cores. Perhaps the most significant thing we have learned about the canyons of the sea is that they are far more complex and far more varied than was once believed.

IGY – Cooperation for the Sake of Science

Before long there will be another International Geophysical Year like the highly successful effort in 1957–58, which added considerably to our understanding of all the earth sciences. In those two years, some 5000 scientists from more than 40 countries made a concerted, worldwide study of the environment, including some significant studies in oceanography and polar sciences.

It was the third such effort. The first had been the International Polar Year of 1882–83. From bases set up in the arctic in those years we learned a great deal about magnetism and weather in the far north and its effect on global weather. The Second International Polar Year (1932–33) gave us new knowledge of radio communication and opened the way for a number of electronic advances, including radar.

In 1957–58 the International Geophysical Year studies in oceanography did much to broaden our understanding of deep currents in the sea as a requisite for long-range weather forecasting. Submarine geophysical studies were made in the eastern and western Atlantic basins and in the central and eastern South Pacific. Sea-level recorders at 30 island stations measured the daily and seasonal fluctuations in sea level and their relationship to other phenomena in the ocean and the atmosphere.

A major effort of the IGY was devoted to the study of interactions of the Southern Ocean and the atmosphere above. Before the IGY

*The oceanographer must be prepared to work under even the most **dire conditions** (below).*

*The **water-sampling bottle** (below right) and a refinement of the bucket **thermometer** (above*

it had not been clear whether the Circumpolar Current involved the entire water body from the surface to the bottom, or whether a deep countercurrent existed below the eastward surface flow. Observations of the distribution of temperature, salinity, oxygen, and other substances proved that the eastward motion persists throughout the entire water column. The total flow, however, seems to consist of a complex of separate streams with fast-moving cores and even some subordinate countercurrents.

From outposts in the far north, task groups set out by ship, dog sled, tractor train, and aircraft to fill in some of the big gaps in our knowledge of the Greenland Icecap and polar Canada. The most spectacular polar projects, however, were in the all-out, all-nation effort in Antarctica, with its 16,000

right) are part of the array of equipment used by ocean scientists to study air-sea interactions.

miles of relatively unknown coastline. Eleven nations—Argentina, Australia, Chile, France, Great Britain, Japan, New Zealand, Norway, the Union of South Africa, the Soviet Union, and the United States—established bases on Antarctica or its offshore islands. Magnetic observations and gravity measurements were made in dozens of locations. A network of meteorological stations made surface weather observations and upper-air observations to 100,000 feet twice daily by balloon with radio transmitters that sent back reports on temperature, pressure, moisture, and wind. Glaciologists drilled holes through the ice to a depth of 1000 feet and more to obtain ice cores and to measure temperature gradations.

The oceanographic and polar studies were only a part of the scientific investigations of the IGY—the most ambitious cooperative study of our environment ever carried out.

Life in the Hadal Zone

Once it was believed that life could not exist at great depths in the sea because of the high pressure and the absence of light. Edward Forbes published a thesis in the early 1840s claiming that approximately 1800 feet was the critical depth. The theory did not hold up for long. The *Challenger* Expedition in the 1870s brought up living organisms from depths almost ten times greater than Forbes's limit. Prince Albert I of Monaco further discredited the theory when he brought up hundreds of bottom-dwelling fish from about 20,000 feet. Almost half a century later, in 1948, the Swedish Deep-Sea Expedition *Albatross* obtained bottom-living organisms from about 25,000 feet. Three years later the record was broken by a Danish expedition aboard the *Galathea*.

The *Galathea* Expedition proved that life exists even at the greatest depths when it dredged animals from as deep as 32,565 feet in the Philippines Trench. This gave support to the notion that so long as there was oxygen and a rain of organic matter from above, life can exist in spite of cold and pressure and the absence of light. The *Galathea* Expedition found a diversity of life consisting of the same basic groups seen in shallower water. They dredged sea anemones, polychaete worms, small crustaceans, snails and clams, and sea cucumbers. Pressure tank tests made recently in laboratories prove that simple invertebrate creatures are practically insensitive to enormous pressure changes. On the contrary, more complex animals suffer damage if they are rapidly compressed to the equivalent of 16,000 feet. The existence of fish much deeper down suggests that cold-blooded vertebrates can adjust to the hadal pressures in the deepest trenches only with time. Highly organized mammals, even if they did not have to breathe gases, would probably be permanently injured at a depth of about 10,000 feet.

The *Galathea* Expedition also made a significant find in the various species of bacteria it brought up in a series of sediment samples from the various trenches and deeps. Just what proportion of the food of deep-sea animals consists of bacteria is not known, but some scientists believe they may constitute an appreciable part.

The discoveries of the *Galathea* were not limited to biology. The expedition recovered materials which are usually supposed to be deposited only in shallow water—fine gray sand, pebbles, cobbles, and land-plant debris.

Deep-sea dwellers (left) have adapted to cold, high pressure, and darkness. They are generally small and either black or transparent. Many species are luminescent.

*A bizarre inhabitant of the hadal zone is the **tripod fish** (right), seen standing on the bottom.*

The New Northwest Passage

Robert E. Peary had gone over the ice to the North Pole in 1902. In 1958 man went beneath the ice to the pole. On August 3 of that year, after two failures, the nuclear submarine *Nautilus* of the U.S. Navy, under Commander William R. Anderson, passed directly beneath the ice at the North Pole. A nuclear submarine like the *Nautilus* seemed to be an ideal choice for the mission. She was capable of descending hundreds of feet and so could pass under the deepest ice formations. Her nuclear-powered engines required no air, thus eliminating the need to surface, a treacherous maneuver in ice-clogged waters. The *Nautilus* had a controlled environment, with the temperature exactly 72° F., regardless of external temperatures. The fuel supply consisted of a small piece of uranium with an energy output equal to tens of thousands of gallons of conventional submarine diesel oil. Because the reactor was small and there was no need to store a lot of fuel, there was room for enlarged dining and recreational areas, as well

as a library, machine shop, and photographic dark room. All these facilities were designed to maintain morale during the long journey beneath the polar ice cap.

This historic voyage began in the Hawaiian Islands. The *Nautilus*, named after Jules Verne's fictional ship in *20,000 Leagues Under the Sea*, then went through the Bering Sea and Strait into the Chukchi Sea. While going through the narrow Bering Strait, the *Nautilus* was sandwiched between ice to the extent that 43 feet of water lay between her keel and the ocean bottom, while only 25 feet separated the ship's conning tower from the ice sheet above. Soon after the submarine entered the Barrow Sea Valley, near the northernmost point of Alaska, it passed under the polar pack ice. It was 1100 miles under the ice to the pole—which took four days—and another 800 miles to the open sea near Spitsbergen.

The objectives of this epic cruise to the North Pole were many and varied. They were to take depth soundings and get water samples whenever possible. They were to observe Arctic currents, measure air and water temperatures, attempt radio communications with the outside world, study ice pack formations, measure light penetration through the ice, and study the physiological effect of the arctic expedition upon the crew of the submarine.

The fathometer, a sonar device, measured distances to the bottom. Half a dozen echo sounders topside recorded distances to the ice above; the echo sounders give off a chirp when the signals bounce back from open water but a dull thud when they come back from striking ice. In addition, television cameras kept a close watch on the ice formations that scudded by like clouds.

Shortly before reaching the pole, they passed over the Lomonosov Ridge, a 9,000-foot mountain range whose existence the Russian scientist Mikhail Lomonosov had first predicted on the basis of geophysical studies of the earth's crust. The temperature at the pole was 32.4° F., and the depth was 2235 fathoms, deeper than had been reported by Ivan Papanin in 1937 and by Admiral Peary in 1909. The ice beneath the pole extended 25 feet beneath the surface. The undersea mountains were described by Commander Anderson as "phenomenally rugged and as grotesque as the craters of the moon."

It is difficult to navigate in high latitudes. Magnetic compass needles swing erratically when they are near the magnetic pole. The success of the *Nautilus* must be attributed in part to the modern gyrocompass.

The *Nautilus* also carried an invaluable instrument called an inertial navigator. Its stable platform always points at the center of the earth. Two sensors measure any changes in acceleration and, hence, changes in direction and speed. A computer is fed with the instrument's signals and gives out information on position. In effect, it shows where the ship is by remembering where it has been and details of the route it has taken.

Another nuclear submarine of the U.S. Navy reached the pole just eight days later. It was the *Skate*, whose captain was Commander James F. Calvert. The route of the *Skate* was from the east. After reaching the pole it paid a visit to Drifting Station Alpha, one of two stations maintained on the ice by civilian scientists and U.S. Air Force personnel during the International Geophysical Year. On its next voyage to the pole in March 1959, the *Skate* broke through the ice to surface, the first ship in history to do so.

The bridge of a submarine stands like a giant monument in the vast emptiness of the frozen polar sea. In 1959, the Skate *became the first submarine to break through the polar ice cap.*

A Hard-Nosed Tanker

In 1968 the largest oil strike in North American history was made off the north coast of Alaska. The immediate question was how to transport the oil to the United States. An economical way would be along much the same route to the east that Fridtjof Nansen's *Fram* had taken in the last decade of the nineteenth century, but surely a ship laden with millions of barrels of oil could not simply drift with the ice. The ship for the job would have to be an icebreaker, and the biggest and most powerful icebreaker ever built.

The project planners decided to give the idea a try and obtained a lease on a 103,400-ton tanker, the *Manhattan*. There are larger tankers in the world, but probably none more powerful. The *Manhattan* was refitted for her special mission. An enormous icebreaker was built into her bow. Instead of being raked at 30°, as smaller icebreakers are, the *Manhattan's* bow began at an 18° angle and bent into the larger angle farther down, allowing more weight to be brought to bear on the ice. In the words of her captain, she proved to be "the best icebreaker ever built." On the first afternoon she was cutting through ice 15 feet thick at speeds up to 12 knots. As she advanced, parties of scientists flew ahead in the ship's helicopter to test the ice for salinity and thickness. (As ice grows older it becomes less saline and stronger.)

Despite all this, the Arctic's large, fast-moving icepacks almost stopped her and forced her to sail an alternate route. But the *Manhattan* proved that it could be done.

The idea of the icebreaking tanker has been abandoned for the time being in favor of the more economical pipeline, but it is not inconceivable that it will be revived. A fleet of such supericebreakers might eventually be able to keep the sea free of ice from Alaska to

the Atlantic. From ships of the *Manhattan* class it is only a small step, at least conceptually, to ships four or five times as big that could cross the Arctic Ocean at will, treating the ice, which averages eight feet in thickness, as a trifle to be easily brushed aside.

But as with all advances in technology, the value of this plan must be weighed against its disadvantages. Such giant vessels are no more subject to disasters and oil spills than any other, but the consequences of such enormous quantities of oil being released to the environment are worth considering. An oil spill would be especially dangerous in the arctic, where the chemical reactions necessary to the breakdown of the oil would be severely impeded by very low temperatures, thus putting a stress on the ecosystem.

Adrift in an Ocean River

During the summer of 1969 a special submersible named the *Ben Franklin* drifted with the Gulf Stream from the east coast of Florida to the chilly waters of Nova Scotia. The Gulf Stream Drift Mission was under the direction of Jacques Piccard and carried five other men to operate the vessel and to make special studies. The ship carried cameras and scientific instruments outside the hull, and every available space inside was given over to recording instruments.

For 30 days the crew floated with the underwater current at depths as great as 2000 feet. Off the coast of the Carolinas, the *Ben Franklin* encountered internal waves (waves between two layers of water with different densities) in the Gulf Stream at a depth of 600 feet, with a vertical motion of about 150 feet every six minutes. The ocean floor here was found to be particularly rough and hazardous, with steep hills over 100 feet high. Most of the time the vessel drifted stably at a depth of between 500 and 700 feet, using no propulsion. She made six vertical trips to the bottom, ranging to 1800 feet, for surveys.

At one point the *Ben Franklin* was forced out of the Gulf Stream by a powerful eddy to the west. After five hours of using power to try to get back into the current, the ship was compelled to surface and to permit her support ship, the *Privateer,* to tow her 50 miles eastward to get back on course.

The voyage reported the almost total absense of fish, plankton, and other forms of life within the warm, turbulent waters of the stream. However, such observation is biased by the fact that alien objects keep mobile creatures far enough away to be out of sight, and by drifting with the water, the sub remained almost without relative motion. One member of the crew concluded that perhaps to study a current, it is advisable *not* to drift with it. Piccard said that he and the crew observed only rare slivers of plankton and an occasional fish attracted by the curious visitor. "Once a small squid came and attached itself to the windowsill above my bed," recalled Piccard, "and we observed each other in complete tranquillity." A swordfish was not so amiable—it attacked a porthole, missing by several inches. "We were probably considered an underwater monster, and our portholes were thought to be eyes," said Piccard. If the voyage was not very meaningful for biological sciences, it brought about other results: the *Ben Franklin* carried all the instruments normally installed on larger, surface oceanographic research vessels and, at the same time, avoided the wind and waves that affect surface ships. The most important discovery the voyage made about the Gulf Stream was a basic one: that it is not simply a broad, steadily drifting "river" as had been commonly thought, but that it consists, in Piccard's words, of "several swirling, colliding, meandering torrents tumbling northward."

In 1969 the **Ben Franklin** *(right) made a study of the Gulf Stream, drifting in its currents for 1650 miles. Above, she is seen aboard her support ship.*

The Drilling Ship

One of the most ambitious projects in the history of ocean research has recently been undertaken. The ship involved is named after the H.M.S. *Challenger*, which had conducted the world's first extensive survey of the oceans: the vessel, the *Glomar Challenger*, was designed by the Global Marine Company and its activities are supported by the National Science Foundation. It began its work in 1968 as part of the Deep Sea Drilling Project and is still at work in 1974.

The Glomar Challenger is a scientific tool especially designed for deepsea drilling. The vessel is 400 feet long and has a 142-foot drilling tower. Among its more interesting innovations is a system of dynamic positioning, which permits the ship to remain practically motionless over the drill site thousands of feet below. A sonar beacon is placed on the ocean floor; its signals

*Fore and aft side thrusters, part of the **Glomar Challenger's dynamic positioning system,** help keep the ship in position over a drilling site.*

are received on board and analyzed by a computer which controls the main propellers and the side thrusters. The side thrusters consist of propellers in circular tunnels that lead from one side of the ship to the other, enabling the ship to move in any direction from forward to sideways. The ship is stabilized to lessen the roll, further reducing the strain on the drill pipe and its operation. The drilling is done through a large hole in the center of the ship. The *Glomar Challenger* is outfitted with the most complete geological laboratory that has ever gone to sea.

Traditional navigational techniques allow positioning with an accuracy approaching one nautical mile, but by using satellite navigation the *Glomar Challenger* is assured of its position to within 0.1 mile. To determine the exact site of drilling, the ship traverses the area using a sub-bottom profiler, depth recorder, and magnetometer. When the ship is in position, acoustic beacons are lowered and the computer takes over the problem of ship immobility. The drill string is then lowered and drilling proceeds as in any land-based oil well. The drill bit excavates a tubular hole leaving within a core of sediment which can be pulled up through the drill pipe and studied. The drill string has been suspended a remarkable 20,000 feet below the ship and drilling has been performed over 4000 feet into the bottom. The oldest ocean sediments recovered in this manner date back 140 million years.

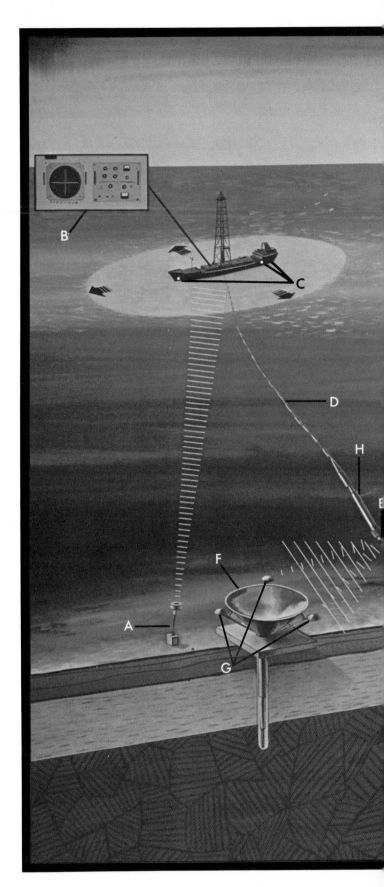

A dynamic positioning and re-entry system enables the Glomar Challenger *to re-enter a hole located far below the surface. Once a* **drilling site** *is selected, a sonar beacon (a) is set on bottom. Its signal is analysed by an onboard computer (b), which takes control of the main propellers and side thrusters (c) and keeps the ship in position. The drilling string (d) is lowered through the hull. A sonar scanner (e) locates the re-entry funnel (f) by detecting its own signal bouncing off sonar reflectors (g). A water jet (h), controlled by the computer, guides the string into the hole.*

Achievements of the *Glomar Challenger*

The Deep Sea Drilling Project (DSDP), of which the *Glomar Challenger* is the principal tool, has been the most elaborate drilling project ever undertaken, and it has made some remarkable discoveries.

One of those discoveries concerns continental drift. The theory of continental drift maintains that a single supercontinent broke up about 200 million years ago. When the supercontinent broke up, fragments of con-

tinental material that were in the breaking zone slowly sank beneath the sea. Drilling by the *Glomar Challenger* has revealed shallow-water-deposited sediments and even indications of dry-land conditions that are overlain with typical deep-sea sediments. By comparing sediment types, some seamounts have also been found to have sunk thousands of feet below the surface of the ocean. A major submarine ridge in the Indian Ocean was shown to have once been a chain of islands with swamps and lagoons. Low-grade coal and peat were found, as well as lagoon sediments and oyster shells.

Drilling results have also confirmed a general increase in the age of the ocean floor from the area of crustal generation at the midocean ridges to the destruction zone in the deep-sea trenches, strongly supporting other geophysical data interpreted in terms of sea-floor spreading and continental drift.

On the smooth floor of the Gulf of Mexico, the *Glomar Challenger* found salt domes which had slowly risen through sedimentary layers in which oil was trapped. This was surprising, because salt domes and oil are usually found in shallow sediment and here the drilling was made at 12,000 feet. The hole was filled with cement to prevent leakage of oil, but the find was made known and will surely encourage future explorations for oil in deep water.

In the Mediterranean, drillings indicated that that sea may have completely dried up about 12 million years ago. This was suggested by the mass of salt deposits and related sediments as well as by the various species of marine animals and plants that were found. Scientists have long puzzled over features on the cliffs descending to the bottom of the Mediterranean that suggest erosion by water. It is now conceivable that great water-falls cascaded from Europe down to the Mediterranean basin thousands of feet below. Imagine the scene created by the rising sea level at the end of an ice age when water began to pour over the Strait of Gibraltar and refill the Mediterranean.

In the *Glomar Challenger's* first voyage to the Antarctic Sea it was discovered that Australia did indeed break away from the polar continent some 50 million years ago and has been drifting northward at a rate of a few inches every year.

*By studying **core samples** (left), geologists aboard the deep sea drilling vessel* Glomar Challenger *have unlocked some of the mysteries surrounding the theory of continental drift.*

Workmen aboard the Glomar Challenger *guide a pipe into position over the drill head (right). A tower supports a **mechanized block** (below) to insert or withdraw the pipe.*

Chapter IV. How Much Life in the Sea?

Of all marine sciences, the study of life in the sea has most intrigued and fascinated the people of the world. The earliest-known naturalists, such as Aristotle or Pliny, devoted considerable attention to the description of marine life-forms. Interest has never abated and yet, even today, all the species of ocean life and their distribution are not known. Rarely a day goes by without some new life-form being discovered.

Biological oceanography is not always so glamorous as it might seem to the young student. Today emphasis is focused upon

---

"Progress in our understanding of the biology of the sea has a bearing on *all* aspects of oceanography, and all aspects of ocean science help us to understand marine life."

---

such vital but unspectacular fields as primary production of the oceans (that is, how much mineral carbon is used to build plant cells through photosynthesis, or, practically, how many billion tons of plants are produced every year in the oceans as a whole) or molecular biology. Mathematics, statistics, and computer programming have become tools as important to the marine biologists as dredges, nets, and microscopes were in the recent past.

Early marine biologists were primarily interested in collecting, describing, and studying organisms. In later years they assessed the relations between marine life and the chemistry of seawater; they studied behavior, interrelationships among organisms, and emergency measures to protect the marine environment from devastation by dangerous pollutants being released into the water.

Progress in our understanding of the biology of the sea has a bearing on *all* aspects of oceanography, and all aspects of ocean science helps us to understand marine life. The study of the biology of the sea can have relevance in unsuspected areas: one cannot construct a river dam in Siberia, Alaska, or Norway, for instance, without knowing about the salmon. If the dam is to be constructed in a river used by salmon, a special kind of ladder must be built to enable the salmon to cross the dam on the way to its spawning grounds, and that ladder cannot be built without knowing how strong a swimmer the salmon is or whether its behavior will allow it to proceed into such a structure.

If we could comprehend how the green turtle is able to swim from the coast of Brazil to Ascension Island in the South Atlantic to lay its eggs and return to precisely the same beach, we might be inspired to imagine breakthroughs in the art of navigation.

The need to provide more food for the world's exploding population is at the root of the increasing interest in marine biological research. Fish population dynamics, studies of growth, of aging; fish tagging; research in fish location on the basis of water temperature, salinity, currents, plankton bloom, or by electronic means—all these efforts are only aimed at chasing fish more efficiently. No serious consideration is being given to long-term management of the ocean's biological resources. The alternative is aquaculture, which requires even more fundamental knowledge than fishing, but in 20 years aquaculture may easily satisfy 30 to 40% of our needs.

Divers from Conshelf II, *our underwater habitat in the Red Sea, study fish in a* **holding tank.**

Chemistry and Life

There can be no satisfactory understanding of conditions of life in the sea without taking into account chemical and physical properties. Three of the most important of these are nutrients insofar as they function as limiting factors; salinity as it affects water mass; and temperature.

In addition to the major elements and trace metals, there are components in seawater of fundamental importance to the growth of phytoplankton, the base of the food chain. These are soluble inorganic phosphate, nitrate, nitrite, ammonium, and hydrated silicon ions. These fertilizers are consumed only in the upper layers of the sea where there is sufficient light for photosynthesis.

It is impractical to analyze every chemical component of seawater. Usually just one element related to salinity is measured. This technique is possible because the principal components of seawater have a ratio that is essentially constant. Salinity is now measured mainly by determining the electrical conductivity of the water.

The most common method of collecting water for chemical analysis uses the Nansen bottle, developed by Fridtjof Nansen. A series of these metal collecting tubes can be strung on a cable and then lowered to whatever depth is desired. They are open at both ends, allowing water to pass through easily as they descend. When they have reached the appropriate depth, a metal weight, or "messenger," is released from above and falls along the cable. When the messenger hits the first bottle, it trips a trigger that tips the bottle over and seals the column of water in it. The reversal of the bottle can also release another messenger to the next Nansen bottle below and so on until all of the bottles on the string have collected samples at various depths. Another bit of information is also obtained as the bottles flip over—the temperature. On each bottle there is a reversing thermometer, which has a loop in the tiny tube filled with mercury. As the thermometer turns upside down the column of mercury is broken, and the mercury in the capillary tube separates from the main bulb or reservoir. It falls to the opposite end where calibrations indicate the temperature when the thermometer was reversed.

Though the reliable Nansen bottle is still in use, it is a time-consuming tool, and ship time is expensive. More and more it is being replaced by modern electrical sensors that continuously transmit all the necessary data to recorders and computers on board.

Nansen bottles on a cable (above). A weighted messenger trips the first bottle (right) and it flips over, sealing a sample of water within it.

Plankton in Balance

Phytoplankton (the plant form of plankton) is the base of the oceanic food chain. It is made up of thousands of species of minute plants that deserve special attention.

Because plankton is the source of all living energy in the sea, considerable effort and research have gone into the development of plankton-sampling techniques.

The basic device to sample plankton is a cone of fine-meshed material mounted on a ring, ending in a bottle or metal cylinder. A major problem of plankton studies has been knowing exactly where the sample of plankton came from. Was it at the surface while the net was being lowered and raised or at some depth below where the tow was intended to be? Was the plankton evenly distributed in the water like mist or were there scattered concentrations like clouds in a blue sky? These questions are vitally important

Bongo nets (above), so called because they resemble huge bongo drums, are able to obtain two samples of plankton from the same mass of water, while taking only a single tow.

*Not all plankton spend their lives floating at the surface. Each of these **planktonic crustacean larvae** (right) may develop into an adult crab.*

*A scuba diver guides a **self-propelled plankton net** (below). This technique permits sampling of a predetermined area at a specific depth.*

to understand the productivity of the waters and the laws governing marine life communities. To some extent these problems were solved by the Longhurst-Hardy continuous plankton recorder. The plankton recorder is a tanklike device with a small aperture leading into a tunnel. A band of gauze travels across the end of the tunnel and is covered by a second band as it leaves the tunnel. Both bands are reeled into a tank of preservative. The "plankton sandwich" is a record of the types of plankton caught and their locations.

The Clarke-Bumpus sampler has been widely used as a collector of plankton. It is a small net designed to be towed at slow speed. It is made of a brass tube, five inches in diameter and about six inches long, to which a silk or nylon net is attached. The sampler is towed at speeds of from one-half to three knots. The instrument has a flow meter that gives a measurement based on the volume of water that has passed through the net.

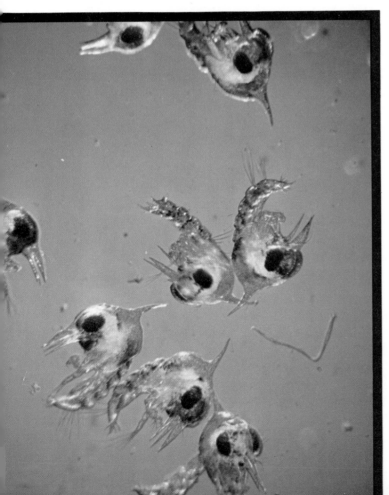

There is doubt about how representative such catches are: the finer the net, the greater the turbulence at the entrance, which swirls some creatures away; even small animals are scared by the waves generated by the contraption, and many are capable of darting away from it. This was proved by cine cameras towed in midwater.

Certain species of plankton are of special value to oceanographers because they are characteristic of particular bodies of water. They are called "indicator species," and they can be used to point out the origin and movement of the water body.

Phytoplankton can be used to fight pollution. Because they are photosynthesizing, they provide oxygen that contributes to the bacterial oxidation of certain pollutants, and they directly remove nitrogen and phosphorous compounds from the sewage.

Periodically a remarkable event occurs that demonstrates the vital importance of plankton to the life around them and also shows how very delicate the balance of the food cycle can be. There are tens of millions of birds, billions of fish, and millions of billions of other organisms that depend for survival on the continual replenishment of nutrients in the Humboldt Current off the western coast of South America. About every seven years, however, wind currents shift and the surface waters are not skimmed off the top of the current. Warmer water from the north forces the current away from the shore. Water rich in nutrients from the deep sea is not brought to the surface. Without the renewed supply of minerals, the plankton, then the fish—principally anchovies—and finally the birds that feed on the fish die. The sea and shores are filled with dead animals. Animals in the sea adjust themselves to a certain set of conditions, and when these conditions are altered and the balance of nature disturbed, catastrophe can follow.

Colored Ribbons and Magnets

As long as large-scale fish farming has not developed, the only acceptable way to exploit a population of fish is to know how many there are and how many there need to be to keep the population at a stable, self-sustaining size. Then it may become possible to appraise how many can be taken without significant harm to the stock. The first matters to clarify are how fast the fish grow and at what points in their life they reproduce. Using this information, we can harvest the fish after their periods of reproduction, thus allowing young to repopulate the group, and soon after the adults have slowed their growth rate, in order to obtain the biggest fish in the shortest amount of time.

Such studies are part of a branch of biological oceanography called population dynam-ics. It includes research concerning the habits of fish as a species, and the ways they are under the influence of their environment. Variations in temperature, salinity, current patterns, and many unknown factors often upset the survival rate of marine life—or suddenly cause population explosions. Research in the area can be difficult and costly. Many unrewarded trials and errors will have to be made before changes in the abundance and availability of fish and the causes of those changes are understood.

At present we know the life histories of only a small number of species. Scientists are trying to estimate the relative abundance of certain fish at different stages in their life. They have tested various methods of making an accurate census of sea animals. Tagging, developed to trace migration patterns, can also help in understanding population dynamics.

*A haddock is **tagged and measured** before release to help predict safe levels of fishing intensity.*

*A tuna gets a **spaghetti tag** to determine migratory patterns and relative abundance when recovered.*

In 1653 Isaac Walton first tagged fish by tying colored ribbons to their tails to help him observe the swimming motions of the fish body and the water flow along the flanks and behind the animal. Now the tags bear a code number and the address of the laboratory to which they should be returned. Scientists have developed many ways of marking fish. One widely-used tag consists of two plastic discs fastened by a metal pin through the back muscles. Other types are the metal clip fastened to the tail fin, mouth, or gill, and the spaghetti tag, a thin piece of plastic with a barbed end that is inserted into the back of the fish. Salmon are often tagged by making an incision in the fish's belly and inserting a strip of plastic. When the fish is processed at the cannery, the tag is discovered. Acoustic tags are being used today and, in the study of the anchovy, magnetic tags.

Tagging can present some problems. It was discovered, for instance, that the bluefin tuna was so fatigued after being brought to a boat for tagging that when it was released it was an easy prey for sharks. The researchers ingeniously devised a method of letting the bluefin tuna tag itself. The tag is a baited hook attached to piano wire. When the tuna strikes, the wire holds just long enough for the hook to sink in, then it breaks.

Biologists hope to get assistance from sport fishermen in tagging operations, and even to divert their interest from their destructive hobby to constructive pasttimes. They are asked to apply tags and release fish and to inform the scientists of the time and place of release. If the fish is later caught, the tag has only to be removed and sent to the address printed on it. The finder receives a small monetary reward for his effort.

*This shark was given a spaghetti tag as part of **an experiment with sound**. Scientists wanted to know if certain sounds would summon the shark. When they saw the tag, they knew the answer.*

Scientific Farming

Recently a British scientist developed a method that drastically shortens the time needed to produce the same successful inbred line. The method is called gynogenesis, and initially it involves the activation of an egg by a sperm cell that is in effect genetically dead. In normal fertilization, the sperm penetrates the egg and the genetic material of both combines so that the embryo contains the same amounts from each parent. In gynogenesis, the chromosomes of the sperm cell are destroyed by exposing the milt to atomic radiation. The chromosomes are killed but not the sperm cell itself, with the result that it is able to penetrate the egg but takes no further part in the development of the embryo. Thus the embryos become "mother" fish. In this way, however, the fish would develop with only one set of chromosomes. The number of chromosomes is doubled by giving the newly fertilized eggs a shock by immersion in very cold water.

So far, gynogenesis has been carried out with plaice, a robust relative of the sole and the flounder, which has a good growth rate. Plaice are easy to work with because their eggs are large, their larvae are relatively easy to feed, and their survival rate is good. Unfortunately people have not yet taken to plaice as a food fish, but tastes can be manipulated by publicity.

Gynogenesis has also been used to produce fish with three sets of chromosomes. These fish grow faster and will be bigger than those with the normal two sets. A fascinating possibility is the production of fish with four or more sets of chromosomes, but in the meantime research is aimed primarily at determining the species that can be manipulated successfully by this method.

*A scientist checks on the growth of an **abalone,** a large marine snail (top left). Abalone meat, known as the "filet mignon of the sea," brings high prices on the world market.*

*A special mixture of fish meal is fed to a thrashing mass of **pompano** (left), one species of ocean fish successfully reared by mariculturists. Pompano are favored for culture because they are a hardy fish, readily adapting to confinement and artificial food.*

*Scientists sort animals from a trawl catch (right) to learn about **food web relationships.** Such data is needed if scientific farming is to work.*

*Scientists prepare to lower a **pneumatic coring machine**, which will be driven into the sediment.*

The Sardine Census

In the early 1900s the sardine began to be fished profitably in the waters off the west coast of the United States. By 1930 it had become the largest single fishery in California. Then, just after World War II, the populations of sardines began to decline rapidly. The decline continued until by 1960 the fish had virtually disappeared. Its near extinction in these waters brought economic disruption up and down the coast of California and the blame was laid on the fishing industry. The very same sequence of events had happened off the coast of Morocco.

Some scientists wondered if overfishing was really the cause of the collapse of the sardine population, and a major effort was undertaken to answer the question. Off the coast of southern California there are a number of basins where sediments are deposited under rare conditions: there is little or no oxygen present in the bottom water. The immediate effect of this fact is the exclusion of burrowing animals from the surface sediments. Thus sediments are allowed to accumulate undisturbed, with the result that they can provide a remarkably complete micropaleontological record. Among the materials constituting these sediments are fish scales. In the thin layers of sediments deposited year by year (varves) in these basins one can find scales of saury anchovies, lampfish, jack mackerel, and sardines. Samples from the sediment layers revealed that over the past 2000 years there has been periodic fluctuations in the populations of sardines. To be precise, the record suggested that there had been 12 major occurrences of sardines over the past 1850 years. It was shown that when sardines are present in abundance they remain in abundance for 20 to 150 years. When their population declines it remains small for an average of 80 years. On the basis of this information it might be predicted that the sardine could be back in California waters by the year 2035. Interestingly (and reassuringly) the sediment studies also showed that when sardine populations were reduced, other open-ocean (pelagic) fishes took their place, most notably the anchovy. They also indicated that the productivity of the sea can only sustain a certain mass of edible fish, and that marine biological resources are far from being inexhaustible.

Through such studies, it would not be possible to obtain a total picture of past populations in the sea. Currents can disturb the sediments. Some marine organisms do not have hard parts that can survive the journey to the bottom when the animals dies, and even the hard parts can be dissolved in the water, on the bottom, or in the sediments. Nevertheless we now know that sediment studies can give synoptic information about distribution and density of some marine populations, and thus predictions about population dynamics can be looked for.

On the other hand, perhaps we are to blame for the demise of the sardine. We don't know. Fish-catch statistics show the characteristic trend in an overfished population. Each year during the decline fewer adults were netted and the boats resorted to obtaining smaller individuals. Such overfishing would reduce the number of adults able to reproduce and eventually bring about the demise of the population in that area.

Sediment corers *may help explain fluctuations in the abundance of various fish.*

Chapter V. The Destructive Forces of the Sea

Recently the U.S. National Park Service gave up a 40-year effort to maintain artificial barriers against waves and storms. The agency had tried to prevent beach erosion by building giant dunes at the Cape Hatteras National Seashore in North Carolina. They finally discovered that their efforts were doing more harm than good. The artificial dunes that had been built to counter the attack of waves and storms had resulted in considerable damage to beaches and had caused an ecological disruption be-

---

"A beach is a dynamic entity. It is constantly in motion just like the sea that created it."

---

cause the force of waves was abruptly blocked by the artificial structures rather than being allowed to dissipate itself in a long sweep across the beaches. It was just one more example of what man has been learning about the sea since he first tried to build on its edge—the sea will sooner or later have its way.

For thousands of years we have tried, usually by a process of trial and error, to construct shoreline defenses, lighthouses, dwellings, harbors, piers, roads, and other facilities that would withstand a common enemy—waves. The case histories of their destruction make fascinating reading. If we had known more about how waves are generated, how and why they break, and how they affect a coastline through erosion and deposition, we might have had more success. Now after 100 years of scientific work, including a concentrated effort over the past 25 years, most of the major features of waves and their causes can be satisfactorily explained in mathematical terms and reproduced in the laboratory.

Only after huge research and development investments have been made is it possible to be successful in building against them.

Of more dramatic consequence than the ordinary movement of water are the great storm surges and the tsunamis. What was learned about their origin, their paths and their speed has been used to build a warning system in the Pacific that has already helped to reduce loss of life or damage to property when they occur.

The sea is not always so spectacular in tearing down our coastline structures. The borers and fouling organisms do their work almost unnoticed until it is too late and a pier collapses. Our efforts to combat them have mostly been restricted to chemical warfare with deterrent paints and creosoted pilings; the very principle of antifouling paints, widely accepted ten years ago, is now seriously questioned; most of them are based on the toxicity of such heavy metals as copper, and thousands of freshly painted ships are daily spreading poisons that will remain in the sea forever.

Another subtle way in which the sea works against our shoreline structures is in the movement of sand along a coast. From the human point of view, either the sand is being carried away from where it is wanted or it is being transported to where it is a nuisance. This fact is responsible for most of our shoreline problems, and it is usually because developers tend to forget that a beach is a dynamic entity. It is constantly in motion just like the sea that created it.

A huge wave releases its destructive fury as it nears shore. Man's efforts to stabilize the surf environment meet with failure in almost every instance.

Sand on the Move

Valuable property is destroyed each year by movements of sand along the coasts by wave-induced currents. We do not yet understand the shoreline processes well enough to predict the consequences of our shoreline development activity. Sandy beaches are a transient feature, sometimes large and at other times almost nonexistent. The amount of sand carried by the littoral current reaches a million tons annually on some coastlines. There is also a seasonal onshore-offshore transport of sand. These drastic shoreline changes occur naturally and any construction on such areas is surely doomed—if not this year, then the following year.

The beach is truly a river of sand being moved in one direction or another along the shoreline. Theoretically it would be possible to dam this "river" in areas where erosion

Violent storms that originate at sea can cause great damage to coastal structures (above) and massive erosion of the shoreline itself (opposite).

was taking place and allow sand to accumulate and fill in the eroded area. As a test, groins were built, extending from the beach outward, and in fact did get an accumulation of sand on the side facing the current. Although successful in many cases, this "solution" often caused other problems. When the flow of sand was stopped, the beach further down on the lee side was deprived of its source of sand and consequently succumbed to wave action and rapid erosion. This of course required another groin, which required another and so on. On the New Jersey shore there are now more than 300 groins protruding into the sea. Another idea that has been tried is the building of breakwaters. The theory was that stopping the waves would stop the longshore current transporting sand along the beach and would result in sand deposits there. This did work, but again beaches "downstream" were subject to erosion because their source of sand had been eliminated. It was also found that so much sand was deposited that the shore soon extended out to the breakwater, necessitating periodic dredging to replace the natural forces of sand drift.

With advanced technology and a solid background of oceanographic knowledge, these problems can be solved. A model must be built of the entire area and of the nearby shores to help adjust empirically the shape and size of the groins, breakwaters, and other defenses. The investment and the decision belong to the state, aided by a reputable research organization, not to a local or municipal authority. The necessary studies involve correlating sand motion with wave action, recording wind and waves, and profiling beaches in detail. Erosion control can be helped by establishing the relationship between the size of the waves and the growth of the sand deposit, called the "spit."

In England experiments were carried out on the movement of pebbles along the coast. Some of the pebbles were given a radioactive coating and then dumped offshore in deep water and nearshore in shallow water. The positions of the pebbles were then checked as often as weather conditions permitted, in order to measure their movement. The pebbles in deep water were measured with Geiger counters mounted on a sled that was towed along the sea bottom. Those that had been put in shallow water were measured with a scintillation counter. With these instruments it was possible to detect those pebbles that had arrived on the beach, even though they might have been buried as much as nine inches below the surface, and to predict the migrations of pebbles.

To save on research, one shortsighted solution proposed recently called for continuously pumping the sand around a groin or a breakwater. The action of the suction pump would restore the flow to the natural river of sand and, hopefully, prevent excessive erosion down the beach. It may very well be that our best solution to these problems is to just avert them—by studying which coastal features are stable, and allowing construction only on those areas we are confident will have some permanence.

The Power of Waves

Waves can be extremely destructive. One of their most spectacular targets has always been the lighthouse, and there have been almost unbelievable stories about their onslaughts against these structures. Some lighthouses have been swept entirely away by the force of waves.

At the mouth of the Columbia River there is a lighthouse that stands several miles out at sea, atop a rock whose nearly vertical walls rise to a ragged surface about 90 feet above the mean low water mark. The entire rock shudders whenever there is a severe storm, and fragments that are torn from the base of the cliff are tossed on top of the rock. Once during an especially severe storm a rock weighing 135 pounds was thrown higher than the light, which is 139 feet above the sea, breaking a hole 20 feet square in the roof of the lightkeeper's house. Another time there was trouble with the foghorn, which is 95 feet above water. When the lightkeeper went to investigate, he found the foghorn was filled with small rocks.

Breakwaters also take the brunt of a violent sea. Blocks of stone weighing as much as eight tons have been detached from a breakwater and moved more than 70 feet. In 1950 during a storm on Lake Michigan, a concrete cap, weighing 2600 tons, was moved three to four feet when pounded by waves about 14 feet high. Engineers later estimated that the wave pressure at that instant must have been about 1680 tons per square inch.

With modern devices such as crystal pressure transducers, electronic amplifiers, and continuously recording apparatus, the data on the destructive potential of waves is increasing and we are better able to defend our shoreline structures.

There are four major kinds of structures designed to combat waves: jetties, breakwaters, seawalls, and dikes. Jetties, usually in pairs, extend into the ocean to confine the flow of water into a narrow zone. A breakwater is built well out from shore to provide a substantial area of quiet water. A seawall separates land from water at the shoreline. A dike is essentially an impermeable breakwater that functions like a dam.

The engineer who designs such structures takes into account the underwater topography and the conditions of potential wave refraction; all available weather charts are consulted to determine dominant wave direction and the probable height of the largest or most damaging waves. The engineer also considers the general configuration of the coast and the proposed locations of piers and wharves. Today nothing important is undertaken without tests with a model of the pro-

posed structure and of the surrounding shores in a ripple tank and then in a larger wave tank. The engineer who designs a pair of jetties must calculate the optimum distance between them in terms of the average volume of water that will flow in and out during each tidal cycle. The best shape of rock to be used must be taken into account. Curiously it has been found that rocks shaped at random by nature serve better than artificial rocks with smooth faces.

By taking into account all of these factors, shoreline structures can be built to withstand the pounding of the sea. Willard Bascom, who wrote the classic work *Waves and Beaches,* summarized the problem best when he wrote that "the engineer must try to understand how the sea acts and learn to take advantage of the geographic and oceanographic conditions so that everything possible is in his favor. Then, on a battleground of his choosing for the short span of human interest, he may be able to hold his own. For the first and most valuable lesson one can learn about the sea is to respect it."

*A **giant wave** smashes against a concrete pier (opposite). Not designed for waves of this height, this structure faces a perilous future.*

***Low tide** exposes the algae-covered remains of a dock (below). Shattered pilings stand as a testament to the destructive forces of wind and wave.*

Waves and Wave Tanks

Like most waves, the waves that generally affect mankind are those generated by the wind. But there are other kinds, rarer and potentially more dangerous, that are commonly called tsunamis, but more properly referred to as seismic sea waves. They begin when faults or deep fractures in the earth's crust, usually deep under the ocean, begin to vibrate as one mass slips against another. The expansive waves that can be generated, traveling in all directions, can reach speeds of more than 600 miles an hour. On the open ocean these waves are not high—in fact, they can hardly be noticed at all—but over shallower areas, they pile up quickly, reaching suddenly to 90 feet or more, sometimes with disastrous results.

The Aleutian earthquake of 1946 (most seismic sea waves originate off Alaska) created waves 55 feet high on the shores of Hawaii, where 150 persons lost their lives and great damage was done to property.

Now that we have a seismic-wave-warning system for the Pacific, accurate predictions of the arrival time of such waves can be made and much loss of life and property prevented. An underwater earthquake occurring anywhere in the Pacific will now set off alarms in the Hawaiian Islands, the west coast of North America, the Fiji Islands, New Zealand, and every other nation, island, and territory using the warning system. The earthquake and the resultant waves themselves are felt by underwater pressure sensors, and the first sea wave arrival is recorded on coastal tide gauges, as sea level recedes many feet before the first wave hits.

The wave tank is the device for bringing waves into the laboratory where they can be studied more closely. A wave tank is simply a glass-sided tank that permits the scientist to observe the inside of a wave. In the sim-

plest kind of tank there is a movable paddle at one end that fits closely but can slide against the walls of the tank. It is hinged at the bottom and driven by a connecting rod which, in turn, is attached to an arm on a variable-speed electric motor. The point of attachment is adjustable so that the wave height can be varied. The period of the waves (the time between two successive crests to pass a fixed point) is adjusted by changing the speed of the motor.

A fairly recent development is the pneumatic wave-maker. Usually several of them are mounted side-by-side along two walls of large square tanks. Waves are created by changing the air pressure beneath a hood so that the water surface rises and falls. As the water surface inside the hood is depressed, the pressure is transmitted to the water immediately on the other side of the partition where the surface is raised. The resulting disturbance then travels the length of the

Dye studies (left), although they give little quantitative information, are one of the best visual aids scientists have for helping them study flow patterns along a coastline.

Plant material is woven together into **giant mats** *(below) that will protect this sandy beach from destructive wave action. The mats are weighted with stones and sunk into position.*

tank. The speed of the blower motor controls the air pressure and thus the amplitude of the waves. The speed of the motor controlling the valve controls duration of pressure and, thereby, wave length.

The U.S. Navy has a wave tank 2773 feet long, 51 feet wide, and 22 feet deep. Its pneumatic wave-maker makes waves up to two feet high and of any length. These wave-generating devices are essential to predicting the consequences of shoreline construction on the natural coastal processes and vice versa. The importance of thorough studies and building scale models cannot be overstressed. In Monaco at La Plage du Larvotto we saw the development of a small portion of coastline proceed without a problem, after models duplicated the current and wave structure of that particular area. Jetties and even the slope of the sand were incorporated into the model and manipulated so their effect would be minimal. It was decided to construct a similar facility directly adjacent to the first. Another study was deemed unnecessary and construction proceeded. Soon after completion the second artificial beach was choked by sediments. There were enough differences in conditions at the two areas that the first model was not applicable in the area a few hundred yards away.

The Battle Against the Borers

Each year, in the United States alone, tiny animals eating their way into ships and waterfront structures cause about half a billion dollars worth of damage. They are the borers and are found everywhere, from Spitzbergen to Tierra del Fuego.

They have been a menace to man ever since he had anything to do with the sea. The "shipworm" and its ravages were well known in ancient times. Since then man has tried to find a successful way to combat them. The Phoenicians coated the hulls of their ships with pitch and, later, with copper sheathing. In the fifth century B.C. a mixture of arsenic and sulfur was used. As early as the third century B.C. the Greeks used lead sheathing.

There are two main types of borers—shipworms and gribbles. The shipworm is not a worm at all, but a relative of the clam. With rows of minute teeth on its shell, it excavates a tunnel in which to make its home. The gribble is a crustacean, and it causes the greatest damage to waterfront structures. Unlike the shipworm, the gribble does not remain in its tunnel but comes and goes. A heavy infestation of them can eat away more than half an inch of wood in a year.

Borers can be introduced into new areas naturally or by mechanical means. Tides and currents can carry the free-swimming larval forms over long distances. Floating wood debris they have already attacked can become spawning areas in new places.

Chemical preservatives are used to treat wood that might be exposed to the borers. Since different borers react differently to different preservatives, considerable research is being done to improve chemicals. Synthetic resins may render the wood less easily penetrable; metal and sheathing is only efficient if it has no breach, no discontinuity. Still no methods are foolproof.

Pilings impregnated with creosote were deemed unnecessary for docks contructed in the Sacramento River during the 1800s because boring organisms did not venture into fresh water. But as man altered the river delta, salt water did invade the area, bringing with it the larvae of borers. Eventually, the docks collapsed.

Another serious and expensive problem to ship and boat owners is the growth of fouling organisms on the bottoms of their vessels. Algae, barnacles, sea squirts, mussels, and a number of other organisms seek just such a surface as a substrate to set up residence. Their presence impedes the flow of water over the ship's bottom, thus reducing the speed the boat can attain. Periodically the surface must be scraped clean at a considerable expense. As with the problem of borers, scientists are seeking chemicals and surface coatings that will repel and deter the larvae of these animals from settling on the hulls, without contributing to poisoning the oceans. Probably the only advantage of harbor pollution is that boring and fouling organisms can not survive the unpleasant environment to infest pilings and ship hulls.

Fouling organisms, shown at right growing on a dock piling, cause about half a billion dollars worth of damage each year, in the United States alone.

A cutaway of a section of piling reveals **teredos,** *or shipworms, (left) that have bored dwellings.*

Intensive research is being carried out in an effort to combat fouling organisms and corrosion. Below are seen two **experimental studies** *in progress.*

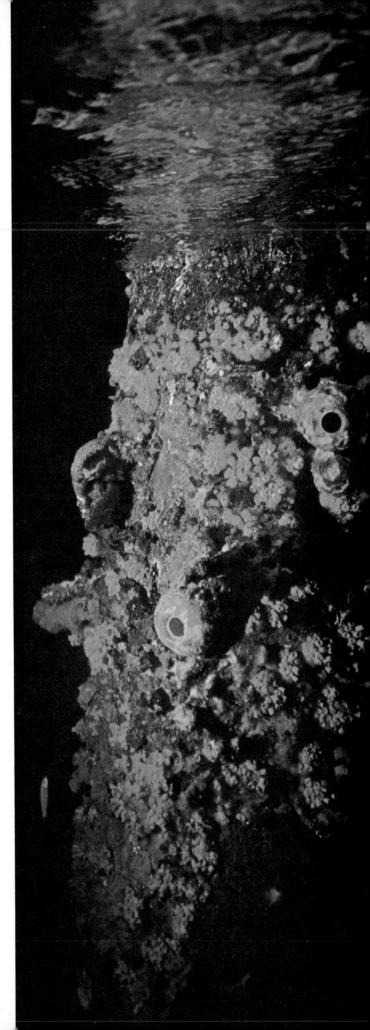

Chapter VI. The Climatic Regulator

In the summer of 1974, 25 to 30 oceanographic research ships from many nations around the world will be stationed in a broad band spanning the tropical Atlantic. Instrumentation buoys will be moored to the sea floor. Weather balloons will be flown to heights of 4000 feet or more. Satellites will transmit almost continuous data on conditions of all sorts in the region, extending from East Africa across the Atlantic to South America and Central America.

A primary objective of the study—the most elaborate and ambitious experiment ever undertaken—is to explore the proposition that clusters of tropical thunderheads play a critical part in lifting humid, energy-laden air from the sea to be swept to other parts of the world. It is this energy, the proposition maintains, that then drives storms far from

---

"We live in a heat machine: the sun is the main source of heat, water is the fluid, and the surface of the ocean is the boiler and the condenser."

---

where their journey began. The existence of the clusters of thunderheads has been made known through the study of photographs from satellites high above the equator.

The experiment will be known as GATE, for GARP Atlantic Tropical Experiment. GARP is a long-term international project whose full title is Global Atmospheric Research Program. The primary purpose of the study is to find the causes of hurricanes and the heat engines of the tropics that control much of the earth's weather. Once it is understood how hurricanes originate, it may be possible to destroy or weaken a hurricane before it wreaks its own destructive force.

Of course, hurricanes are only the most spectacular manifestations of our weather, and many other far more subtle operations are worth at least as much attention. The oceans in their daily activities serve as the regulators of the earth's climate by capturing and releasing vast quantities of energy.

The sea's capacity to store heat has a steadying influence on the temperature of the world. On the surface of the earth, heat is absorbed by the sea only by the uppermost water layers. These layers absorb a great amount of heat from the sun during the day but give it up only very slowly during the night. These characteristics of water help to moderate the climate of our planet and steady temperatures on land. Sea breezes during the day cool the land, while warm currents such as the Gulf Stream carry tropical warmth to colder countries in the north. These currents profoundly temper frigid climates at the poles.

We live in a heat machine: the sun is the main source of heat, water is the fluid, and the surface of the ocean is both the boiler and the condenser. Evaporation and condensation transfer enormous quantities of energy, greater than thousands of hydrogen bombs. In a distant future, when we will have acquired a detailed understanding of the processes involved, through many such global actions as GATE, it will be possible for man carefully to manipulate evaporation and modify weather at will anywhere in the world. This would put an end to calamities like drought, floods, and famines.

*A **waterspout,** reaching downward in the shape of a funnel from a dark **storm cloud,** twists its violent way along the surface of a harbor, menacing boats that might lie in its path of destruction.*

A Bathythermograph, lowered and hoisted while the ship is underway, traces a continuous curve showing temperature in relation to depth. The profile is etched on a smoked glass slide.

Climate and the Ocean

In all parts of the world, climates are shaped by ocean currents and other great movements of water. Water has an exceptional capacity both to store heat and to transport it from one place to another. Some of the principal currents that influence climate in various parts of the world are the Benguela Current, bringing cold water up the west side of Africa and keeping a long strip of that coast relatively cool and foggy; the Peru Current, bringing water of antarctic origin almost to the equator; and the Gulf Stream, with its tempering effect on the climate of northwest Europe.

Our knowledge of such currents comes to us not so much through the use of current meters and drift bottles as through the study of salinities, oxygen content, plankton, and temperatures. From such physical observations we can learn where a particular current comes from and where it is going.

Temperatures were first measured with a bucket thermometer. Later more elaborate devices were used. There are two basic types of thermometer used in conjunction with the Nansen bottles. One is protected from pressure and the other is unprotected. They are used simultaneously, and the difference in temperature reading also indicates the actual depth at which the measurement was made. The bathythermograph, or BT, is a continuous recorder of temperature versus depth. It contains a bellows that contracts with increased pressure as the instrument descends in the sea and a Bourdon element that responds to temperature changes. This instrument at the end of a light cable can be lowered into the ocean from a moving ship. During its descent a temperature record is plotted against depth on a small glass plate.

For depths to 600 feet, it can be lowered while cruising at 12 knots to bring back the

As the BT descends, a pressure-sensitive element forces the slide to move in one direction while the stylus records temperature in the other direction. The result is a curve of temperature in relation to pressure (depth) that is analyzed in a special reader.

greatest volume of ocean temperature data now available. Expendable BTs are even dropped into the ocean from the air. The importance of studying temperature in relation to depth lies in the layered structure of the sea. The currents moving surface water do not extend to the bottom. Below them lie other water masses, some are stationary, and some flow in exactly the opposite direction and are known as countercurrents.

With the progress that we have made in the science of ocean dynamics, the influence of very small temperature differences becomes noticeable. For some studies, temperatures are measured with an accuracy of 1/100th or even of 1/1000th of a degree. Temperature measurements are made today by electrical methods—by measuring the variation of resistance of an electrical conductor, for example, or by the variation of the vibration frequency of a quartz crystal.

Some of these instruments can be overwhelming: the famous "thermistor chain" used by Woods Hole Oceanographic Institution was an enormous reel on a ship's fantail, storing hundreds of feet of streamlined chain, housing hundreds of thermistor sensors, all connected electrically to recorders on deck. Once the chain was unreeled and hanging under the hull, the ship traveling at speeds up to eight knots recorded simultaneously in great detail and with great accuracy the water temperature at all levels for thousands of miles and as deep as 600 feet and even deeper. Such an abundance of data can only be processed by computers.

NIMBUS

A giant step in applied meteorology came with NIMBUS, the weather satellite program of the National Aeronautics and Space Administration. NIMBUS has already had a wide variety of assignments—making the first vertical temperature readings from space through clouds, monitoring a mysterious disappearing current off the west coast of South America, and thermally mapping the earth's surface.

Many parts of the world are under cloud cover more than half the time. With the instruments that NIMBUS carries, these clouds can be penetrated and precise temperature readings obtained, a great help to meteorologists in improving weather forecasts over long periods. NIMBUS's instruments also differentiate between old and new ice in the arctic and antarctic regions, another big help to weather experts. Heavy rain clouds, which may be otherwise indistin-

ERTS, the Earth Resources Technology Satellite, collects, among other things, global data on hydrology, ocean currents, and air and water pollution.

The NIMBUS satellite is another of man's mechanical eyes circling the globe, carrying with it advanced sensors for weather research.

guishable from cirrus or stratus clouds containing little water, are also being identified.

A related job of NIMBUS is helping to make daily charts of the Gulf Stream. By knowing the stream's specific location each day, southbound ships will try to avoid it, while northbound ships will ride in the stream receiving a few extra "bonus" knots per hour. Substantial savings will thus be realized by shipping at a global scale.

The periodic damage caused by El Niño in the Humboldt Current near Peru will be lessened, it is hoped, by observations of sea surface temperature, vertical atmospheric temperature, and cloud distribution in the area, all carried out by NIMBUS.

Weathermen are predicting that as a result of satellite pictures showing the location of ice masses, shipping operations may soon be extended several months—perhaps even through the whole six-month polar night.

NIMBUS has also revealed that the polar ice caps have boundaries that are not accurately drawn on world atlases. The pack lines at both poles are not smooth around the ice edge as shown in the atlases, but consist of a great many indentations. If a navigator used a standard atlas to sail a ship into the area, he would be surprised to find that there were many coves and channels extending into the icepack itself. Images from NIMBUS have established their existence.

NIMBUS images are also showing that the polar regions undergo large-scale changes in a short time. In particular, the boundaries between the multiyear icepack centered around the North Pole and the large areas of the first-year ice are found to vary significantly with one freezing season. Even greater changes in the ice cover occur in Antarctica.

The weather satellite has also been measuring rainfall over the oceans on a daily basis. Before NIMBUS, weathermen had no adequate way to monitor ocean rainfall on a global scale. Knowledge of its rate and extent can give us a good idea of how much heat energy is being released into the atmosphere. In turn, knowing rainfall rate and the amount of heat released will greatly aid in reaching the goal of long-range weather forecasting and improving short-term forecasts of hurricanes so that lives and property along coasts can be protected from the onslaught of storms.

Meteorologists *aboard a research vessel prepare to launch a weather balloon. Data is monitored onboard ship and relayed to onshore weather stations.*

The Acrobats Among Ships

Probably the most unusual looking vessel used for oceanographic research is FLIP, the Floating Instrument Platform operated by the Scripps Institution of Oceanography. FLIP assumes the vertical position in which it carries out its undersea investigations by flooding its long stern section. When the ship turns upright, the bow remains out of water. It is in the bow that the research laboratories, living quarters, and the engine room are located. Two watertight cylindrical tubes allow the crew to descend to 150 feet below the water level. To restore the ship to a horizontal position, compressed air forces the water out of the section that is submerged.

There had long been a need for just such a stable open-sea working platform. In its vertical position FLIP, because of its size, is unaffected by the motion of the sea surface: it does not move vertically under the influence of ordinary waves and swell. This makes it an excellent platform for carrying out undersea sound measurements.

FLIP played a major role in the Barbados Oceanographic and Meteorological Experiment (BOMEX). An area of the Atlantic, east of Barbados and north of the equator, is called the Doldrums. It is of special interest to scientists because hurricanes spawn there. BOMEX was a combined air-sea investigation to determine the thermal interaction between the air and the water.

Among other studies, FLIP recorded the swell of storms occurring almost half the world away. It has also served as a quiet,

Mysterious Island, the first manned oceano-graphic research buoy, anchored in deep sea, stands like a giant top on the face of the ocean (above). Its design provides for a helicopter platform and an elevator inside its 180-foot stem.

Flip, the 355-foot-long Floating Instrument Plat-form operated by the Scripps Institute of Ocean-ography, is towed out to sea (below). When in a vertical position (opposite) about 300 feet of FLIP's cylindrical hull is submerged. In this position FLIP offers a very stable platform for research. With the bulk of the ship well below the surface, waves hardly have any effect on it.

drifting platform from which studies in sound propagation have been carried out, as well as observations of ocean-bottom and biological noises. It has done significant work on heat transfer from the earth to the oceans.

At the same time as FLIP was tested, my research and development unit in the Medi-terranean, CEMA, built the first stable float-ing laboratory anchored in deep water. *L'Ile Mystérieuse* (Mysterious Island), named after Jules Verne's famous book, was an "un-flippable FLIP." It remained seven years anchored in 8900 feet of water, halfway be-tween Nice and Corsica, with renewed teams of six scientists. It studied biology, produc-tivity, evaporation, and currents.

Another unusual research vessel and one that, like FLIP, is tilted to a vertical posi-tion, is SPAR—the Seagoing Platform for Acoustic Research. SPAR is unmanned and operated by two nearby auxiliary vessels, attached to SPAR by power and data trans-mission cables. SPAR's primary purpose is the study of submarine passive listening phenomena and detection ranges under all oceanic conditions. Of particular interest in its studies are the acoustics encountered along vertical oceanic interfaces, such as the temperature discontinuities along ocean cur-rents like the Gulf Stream.

Chapter VII. Location of Minerals

Seawater contains at least traces of all the important elements known to man, including certain minerals that are many times heavier than water itself. A cubic mile of seawater contains 6 million tons of magnesium, a very versatile lightweight metal, and huge quantities of other chemicals such as calcium, sulfur, sodium, chlorine, potassium, bromine, iodine and 38 pounds of gold. The sea also holds dissolved salts of silver, aluminum, iron, nickel, chromium, radium,

---

"Patterns of mineral distribution are emerging from the new concept of plate tectonics."

---

and uranium. A cubic mile of ocean contains enough salt to supply the world's needs for about nine years.

The ocean's mineral deposits include hundreds of other valuables, including solid metals, crystals, fluids, and gases. The most abundant pure metal is manganese, and magnesium metal has been industrially produced from seawater on a large scale since 1941. But the soil under the sea also contains accumulations of other metals such as copper, nickel, and cobalt.

Shellfish, calcareous algae, and coral continually extract calcium from the sea and make it available to man in their shells or in limestone (calcium carbonate) in old reef beds.

Cell tissues of certain marine plants and animals have become highly efficient in concentrating rare compounds, such as the iodine salts found in seaweeds. Sea squirts somehow extract vanadium from salt water, yet scientists have not been able to find this trace element by analysis of the water itself. Other shellfish and seaweeds concentrate such elements as copper, cobalt, and silicon.

A better understanding of the undersea distribution of mineral deposits in both space and time has derived only quite recently from a new conception of the earth that is as dramatic and as significant as the Copernican revolution that taught man to regard his universe as sun-centered rather than earth-centered. It represents the earth not as *static* but as *dynamic*. The whole earth is constantly in a process of change. The basis of this new conceptual framework is plate tectonics—the movements of the rigid outer shell of the earth (the lithosphere), which is segmented into six primary slabs, or plates, and behaves as if it were floating on an underlying plastic layer.

The Red Sea displays a good example of the drifting of continents. The plate boundaries between Africa and Eurasia provide a natural laboratory for the study of mineral formation processes. Recently the richest submarine metallic sulfide deposits known were found in three small basins along the center of that sea.

Plate tectonics is shedding new light on the formation and deposition of oil, as well. Certain plate boundaries create conditions that form accumulations in small ocean basins and deep-sea trenches marginal to continents. Another kind of plate boundary creates conditions favoring the development of oil deposits extending from the continental shelf into the deep-ocean basin under the continental rise.

Patterns of mineral distribution that are emerging from the new concept of plate tectonics will certainly be a great help in guiding man's search for new resources.

*Rich in silicates, the skeletons of one-celled animals called **radiolarians** are important constituents of certain deep-sea sediments.*

Science in the Search for Oil

Important roles in the search for oil are played by the micropaleontologist and the geophysicist. The micropaleontologist is helped by the scanning electron microscope, which can magnify tiny marine organisms up to 140,000 times their actual size.

Electron microscopy has helped paleontologists *identify marine organisms associated with oil-bearing strata. These foraminifera, as well as radiolarians and coccolithophores, are good index fossils.*

Mapping the Sea Floor

*This instrument, called a **FISH**, will be towed underwater to record bottom topography.*

Inaccessible and distant, the floor of the deep sea has been mapped with less certainty than the moon. Beneath the expanses of the open sea lie uncharted mountains, plateaus, great faults, and mineral-rich sediments.

A study of these features will allow us to understand better and make optimum use of the sea. An artillery of observation and measuring devices have been employed to probe the secrets hidden miles below the surface. Corers retrieve a few feet of sediments; bottom grabs bite out mouthfuls of ooze; dredges scoop up assemblages of rocks and surface sediments; and cameras provide quick glimpses of the bottom. Less directly, acoustic devices explore the topography and penetrate the surface with bursts of intense sound showing what lies below the path illuminated by the sound.

One piece of advanced armory is the FISH —a multimedia device designed to do geophysical mapping of the sea floor. It is suspended from a ship above and relays its information back by cable, thus avoiding most of the ship's rolling and pitching, most of its noises and the sound-scattering effect of the surface waters full of tiny organisms. In mapping, precise positioning is important and the FISH is able to interrogate sound transponders nearby on the bottom for orientation. It also possesses acoustic devices which relay the instrument's distance above the bottom and distance below the sea surface. Its sensors include magnetometers which detect variations in the earth's magnetic field and the magnetic characteristics of the crustal rocks immediately below. This

86

information is useful to those who study sea-floor spreading because they can date the speed of the crustal movement by correlating the magnetic properties in the rocks to known dates of pole reversals.

Some bottoms reflect sound very well, while others send back a very poor signal. The mapping of rough bottoms is important to navies in locating the possible sites where submarines could lie undetected by surface sonar systems. TV cameras on the FISH enable scientists above to observe the details of bottom topography and at the same time correlate it to their acoustic profiles.

Perhaps the most interesting information revealed by such deep observation devices is that related to manganese nodules. Deep-sea photographs show vast beds of roundish nodules made up of alternate layers of manganese oxide and iron hydroxide. The growth rate of these nodules is not positively known, but some studies indicate it may be as slow as two to four millimeters per million years. Some beds are remarkably uniform in the distribution and size. Two of the most puzzling questions are why they do not become imbedded in the sediments and why they are in some cases perfectly round. They consistently lie on top of the sediments and are not flattened on the bottom. This would indicate that they are rolled often by currents exposing all sides to equal growth, but we do not see depressed moats where the sediments would be eroded by flowing currents. Scientists are also perplexed by the source of manganese. The nodules are found beneath areas of low-oceanic productivity and not along the mid-oceanic ridges or near the submarine volcanoes, where one would have expected them.

*This **echosounder** can be towed behind a ship where its streamlined design reduces drag and limits the amount of surface noise.*

*A cloud of sediment rises as a **bottom grab** comes to rest (above). As the cable is retrieved, steel jaws trap a sample of the upper sediment layers.*

Bringing Up the Bottom

There was a day when old mops were used to obtain samples from the bottom of the ocean. A certain Captain Carver of the ship *Porcupine* is said to have had the rope mops that the sailors used to swab the decks attached to the ships' dredge, and it resulted in a substantial increase in the catch of biological specimens.

Dredging is still important in the study of the ocean bottom, and a number of forms of dredges are used. The most common kind is the bag hung behind a rectangular or triangular iron frame. Dredges are generally made so that they will function equally well no matter which side lands on the bottom, and the lips of the dredge are usually beveled to scrape the upper layers. For special purposes, the dredges may be equipped with teeth to plow the bottom and stir up the burrowing organisms such as mollusks and worms. There are also "bottom grabs" of the clam-shell-bucket type and the Ekman and Petersen dredges. These fall on the bottom in an open position. When a "messenger" of brass

tubing is sent down the line and releases a catch, powerful springs close the mouth of the bucket, causing it to bite into the bottom, taking whatever of the upper layer is between its jaws. The drawback to the dredge as a bottom sampler is that the materials it brings up are mixed and there is no indication of their natural distribution.

To obtain the subsurface material in the natural arrangement of layers, a core sampler is used—a steel tube that is driven into the bottom and brings up a sample. The tube is usually lined with a smaller plastic tube that can be withdrawn and stored for future study without disturbing the original layering of the deposit it contains. The main tube can be driven into the bottom by allowing a heavy weight to drop on it after it has been positioned—like driving it into the earth with a sledgehammer. A more recent and more effective method is the application of force from an explosive charge, such as the "Piggott gun" attached to the top of the tube. The tube is literally "shot" into the bottom.

By examining these cores of sediment we are able to study the past history of the earth.

Cores from the bed of the ocean have already helped us know the exact dates of the ice ages and to follow the changes that have taken place in the temperatures of the sea. There are probably only a few places where really undisturbed sediments are to be found. One of these is the flat tops of those seamounts that are out of reach of turbidity currents. The difficulty with this is that many of those small flat-topped hills or mountains are not nearly as old as the oceans. They are volcanoes that have erupted from the sea floor fairly recently in geological history. They are not, therefore, covered by the entire thickness of sediments that would reveal the complete life story of the seas.

*Scientists empty a **bottom dredge** onto the deck of a ship. The bag is made of metal chains to withstand scraping on the bottom.*

*Fish caught in a **trawl** spill into a pan. Towed behind the boat, it moves over the bottom, catching fish and other animals of the benthos.*

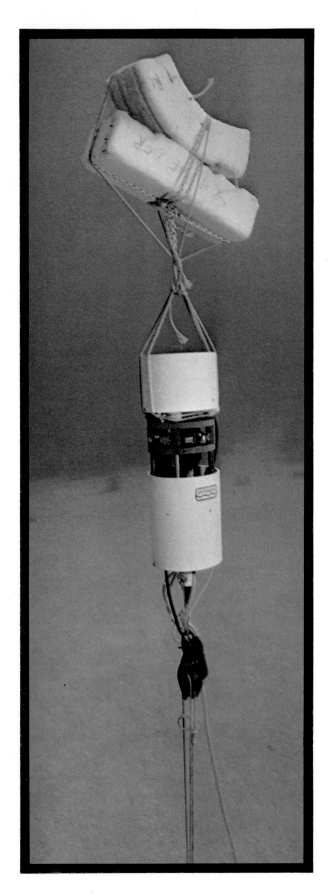

The Red Sea Brines

One of the most interesting discoveries within the last few decades has been the location of extremely hot pools of very salty water sitting in three small basins at the bottom of the Red Sea. The tale begins with the Swedish *Albatross* expedition which was investigating the Red Sea in 1947-48. Their instruments revealed water temperatures of 76° F. instead of an expected reading of 68–72° F. and a salinity of 45 parts per thousand instead of 40 parts per thousand. The scientists were surprised but recorded the measurements data and checked the instruments to see if they were functioning properly. The equipment was recording correctly. Subsequently the *Atlantis* and *Atlantis II* of Woods Hole Oceanographic Institution and the *Discovery*, a British research vessel, confirmed the Swedish findings. Their data indicated temperatures of 111° F. and salinities up to 255 parts per thousand. Something truly remarkable had been discovered. The waters were found to contain up to 50,000 times the concentration of some metals over that of normal seawater. And sediment cores taken below the brines had the same mineral content as the water immediately above. Such metals as iron, manganese, zinc, lead, copper, silver, and gold in the water and in the sediments were valued in 1969 at over two billion dollars. Even before the hurdles of mining them can be faced, laws of the sea must decide which adjacent countries have the right to claim these riches.

An early explanation for the existence of the hot pools in the Red Sea was that since evaporation is greater than precipitation in the

***Modern oceanographers** use a multitude of elaborate electronic devices. The sensing device at left, anchored to the bottom, measures various physical parameters including temperature, pH (used to indicate acidity or alkalinity), and salinity.*

origin, and their salt and metal contents suggested the route that they traveled.

Several hundred miles away from the pools, at the southern end of the Red Sea, the water matches the brines in isotope content. If the water from this area is the source of the brines, it must have traveled through the rocks below the sea floor before reaching its destination. In doing so it could easily have acquired the salts and metals by leaching them from the rocks through which it percolated. If this is indeed what happened, it would have taken a few thousand years for the water to make its journey. Further evidence to support the hypothesis came from samples taken by bottom corers.

Nansen bottles line the wall of this ship's laboratory (below). Analysis of water samples from the depths of the Red Sea led to the discovery of mineral-rich brine pools on the bottom.

Perched out over the waves, an oceanographer prepares a *Nansen bottle* (above). He checks its twin thermometers, as well as the "messenger" attached under it, before its descent.

sea, the saltier, denser surface water that forms by evaporation in coastal shallows drains continually into the sea's central deeps. The investigation showed this was not the case. The chemical composition of the hot brines proves that they were not formed by evaporation. Studies of the isotope content of the brines also ruled out fresh water as a source. It further ruled out "juvenile water"—previously uncirculated water that is brought to the surface by volcanic sources. The brines must be derived from the Red Sea itself. The source could not, however, simply be the water over the deeps. Where, then, does it come from? Again, the isotope content of the brines offered a clue to their

Chapter VIII. Vibrations for Science

The story of underwater sound had a humble beginning in a freshwater lake but subsequently developed into the single most useful tool man has in the sea. In the early nineteenth century two scientists submerged a church bell in Lake Geneva and measured the time it took for the sound to cross the lake. Sound was found to travel four times faster in water than in air. By the early 1900s the Submarine Signal Company was formed to develop a system of underwater warning bells located near dangerous obstacles and rocks. Their sounds would travel

---

Underwater acoustics is the single most useful tool man has for studying the sea.

---

farther than the airborne signals of foghorns and be picked up by receivers on ships, warning them of dangers ahead.

The French naval engineer Marty was the first to use echosounding techniques to measure the depths of the ocean: the sound source was a bullet fired into the water, and an instrument recorded both the impact and the echo. Two other Frenchmen, Langevin and Florinson, built the first echosounder using magnetostriction as a sound source.

By 1922 a continuous acoustic profile across the Atlantic was made.

Echosounding was a revolution. It did away with all the time consuming inconveniences of cable sounding: stopping the ship, lowering a cable thousands of feet to the bottom, noting how many feet were played out, and then reeling it all in. With Sonic Depth Finders the ship never needed to stop. During World War II, mainly in Britain, high-power acoustic transmitters were developed which could be beamed like light and swept

around to "observe" named "sonars." The systems did not provide clear windows through the sea though because of variations in the water's temperature and density.

Sounds were affected by the physical conditions within the sea. For example, sound is bent (refracted) by layers of water at different temperatures. The boundary of two such layers of water is termed a thermocline, and as sound travels across the thermocline obliquely, it may be split, with one beam bouncing back to the surface and the other bending downward. Between them, there is a shadow zone which can conceal a submarine. Various types of bottoms reflected sound differently, with rough bottoms such as canyons distorting the echo to such an extent that a submarine could also hide there.

At the beginning of World War II sonar operators were dumbfounded by animal sounds. Rumbles, grunts, thumps, moans, shrieks and whistles transformed a world of silence into a noisy living sea.

Acoustic research soon zeroed in on the Deep Scattering Layer, an acoustical false bottom that scattered sound and was eventually found to be made up of organisms.

By the end of World War II underwater sound had expanded from the limits of communication and warfare to one of the most useful tools of ocean scientists. Sound allowed them to study water temperature and depth, bottom and subbottom features, and populations of life in the deep sea. It also has stimulated extensive research on dolphins, the living sonars of the sea.

*Biologists want to know the role **sound** plays in the lives of marine animals. Here sharks are attracted to sounds from an underwater transducer.*

*Sound detecting devices make it difficult for **submarines** (above) to travel unnoticed.*

*A **parabolic acoustic device** (right) enables scientists to learn more about sound in the sea.*

Hide and Seek with Sound

A submarine can be found if it can be heard. Today, with the great reliance that the major powers are putting on submarines as weapons, knowledge of the behavior of underwater sound and technology for making use of its properties have without doubt become of paramount importance.

We have seen that a submarine can successfully hide from sonar and other acoustic spotting devices either in chaotic bottoms such as narrow canyons, or, when the submarine is navigating on the high seas, it can hide in the so-called shadow zone. This is an area where relatively little sound can penetrate. It occurs in the upper parts of the ocean where a layer providing high sound velocity overlies a layer that affords low sound velocity. If the object to be detected is in the layer of high sound velocity, its sound will be refracted upward and downward into areas of lower sound velocity, thus producing the shadow zone. It can be very difficult to detect a submarine here.

On the other hand, there is a layer in the oceans through which sound travels very well and can be detected from anywhere in the world. It is the SOFAR (Sound Fixing and Ranging) channel. SOFAR was discovered by Maurice Ewing, who in the 1940s foresaw the importance of underwater sound in submarine warfare. SOFAR depends on the fact that at certain depths in the oceans, roughly between 2000 and 4000 feet, there is a layer in which sound waves travel at a minimum velocity. Sound waves generated within this layer cannot leave it because they are reflected back by the water layers above and below, both of which have different acoustic properties. By triangulation from several listening stations, the source of the sound can be determined to within a mile. One of the practical uses of SOFAR is

to locate fliers downed at sea if they are equipped with the proper emergency signaling equipment. An explosive charge detonated in the SOFAR channel has traveled as far as 25,000 kilometers.

Inventions for use in war often find peacetime applications. Certainly the outstanding example of this is nuclear power, which once wreaked the most terrible devastation man had ever known and which now holds the promise of becoming an indispensable source of energy. The SOFAR channel is not nearly so spectacular, but it, too, is finding several listening stations, the source of the its applications in time of peace. It is being used to study deep-ocean movements. Floats are being set in the channel and are being tracked by the signals they emit. The hydrophones that pick up those sounds can be located many miles away. Hitherto there have been very few precise observations of average deep-water motions because of the limited duration of experiments requiring the tracking of floats by surface ships.

A newly installed **rubber sonar dome** bulges from the bow of a U.S. Navy frigate (below). Sonar is a great aid in the detection of submarines and other objects lurking below the surface.

Navy personnel stand by as a **one-ton variable depth sonar unit** is raised to the surface (above). The equipment is designed to detect submarines hiding beneath thermal barriers in the sea.

Making Use of Echoes

Acoustic techniques have become one of the most efficient and versatile tools for scientific study of the oceans. For accurate measurements, the speed of sound, as well as the physical characteristics of seawater, must be known with precision.

The speed of sound in seawater is about four-and-one-half times its speed in the air. It is dependent on the temperature, salinity, and pressure of the water and, of these, temperature is the most critical. The speed of sound increases with increased salinity or increased depth (pressure).

A refraction or a reflection of sound leads to the creation of shadow zones—where sound is greatly reduced and which can provide a protective shield for submarines that want to avoid detection—and sound channels (or rather, layers) where sound is concentrated and travels extremely long distances. There are often two of these layers, the lower one being the most efficient.

Other distinctive layers, which we have described in these books in their various aspects, are the deep scattering layers (DSL), sound-reflecting layers caused by the presence of marine organisms rising towards the surface at night and descending at sunrise.

Measuring of the ocean depths by dropping a line to the bottom and hauling it back up began as early as 2000 B.C. with the Egyptians. Echo sounding, as it has been perfected today, has enabled research ships to start a detailed global mapping of the ocean floor. The first edition of a world map was prepared by Prince Albert I of Monaco at the Oceanographic Institute. These maps are updated today by the International Hydrographic Bureau, also in Monaco, with the help of all the hydrographic offices of the world. Today, it is possible to make acoustic soundings with an accuracy of a few inches, if the velocity of sound has been acoustically determinted and the tides are well known. The main inaccuracies in such maps originates in the fact that the ship's position is not as accurately measured. Satellite navigation systems do not provide sufficient accuracy. Very expensive radio-navigation systems are the only ones today capable of locating the ship within one yard.

Echo sounding is also used to help marine geologists identify undersea rocks and help locate important mineral deposits. An underwater explosion is detonated to generate a series of sound waves that fan out to the ocean bottom and to the rocks beneath the sediment layer. The echoes vary as the rocks that reflect them vary. Sound waves travel faster through some rocks than through others. Thus, by comparing the acoustic properties of known rocks with those of the sea bottom, geologists can identify the rocks that make up the sea floor.

The echo sounder has proved of considerable value, too, in locating schools of fish, especially herring, at depths to more than 500 feet. The echo sounder not only tells the depth at which the school is traveling but also gives information on its size, density, and upper and lower limits.

*An explosive charge is detonated in **an experiment with underwater acoustics**. A detonation in the SOFAR channel can be detected at coastal stations and used to determine the position of the ship.*

*A **SOFAR float** is recovered by ocean scientists aboard the O.S.S.* Researcher. *Underwater acoustical equipment is helping scientists track the movements of deep-water masses.*

Chapter IX. Assessing Man's Impact

As societies and nations grow and become more complex, the environment becomes increasingly affected by our activities. We are upsetting and eliminating natural land populations through agriculture, mining, and growth of our cities. We have been over-exploiting the sea and dumping great quantities of waste into it as though it were both an inexhaustible bounty and a sewer. The coastlines are most dramatically affected through the development of harbors, coastal cities, and landfills, eliminating the natural life of bays and estuaries.

The only hope we have of saving the sea is in sufficiently understanding it to predict the consequences of our activities. Presently we merely make educated guesses and hope that our activities will not devastate entire ecosystems. A good example of such educated guesses is the environmental impact studies that many states insist must precede any development. The developer is required to submit a statement that, among other

> **"The only hope we have
> of saving the sea is
> in an understanding of it."**

things, describes how his program will affect the coastal ecosystem. The idea is good but the goal is presently impossible to achieve. Since the investigators are paid by the developer, the most serious consequences are sometimes treated lightly. Damage is not proportional to the surface eliminated. But that is only a superficial guess. What about the animals that are transient visitors to the area or only breed there? Because they are part of other ecosystems, anything that affects them will necessarily affect those other areas. The concept of food chains is obso-

lete; intricate webs best define feeding relationships. But feeding is only one of many ways individuals interact. Some depend on others for shelter, a place to breed or a surface to lay eggs on, and even as a service for symbiotic cleaning activities. There are so many interrelationships of marine animals and plants that it is impossible with our present state of knowledge to define which other organisms will be affected by the loss of even one species. Because time is supposed to be money (also a very obsolete concept!) the length of time allotted to acquiring data is extremely short. Shoreline features, water conditions such as currents, and sizes of populations are not static but are constantly changing. It is totally impossible for a six-month study to determine anything but some superficial information on what happened during even *that* period of time. Probably no information exists on fluctuations of even the recent past. At present we are incompetent to judge the consequences of our actions and, once an imbalance or problem does occur, we are most often unable to determine the exact cause of it. Monitoring the sea must begin immediately so we can assess our impact.

We must know what is there before we can understand how it will be upset. We must also study how the system works—how plants and animals interact, how the environment affects them, and how they influence the environment. To date, such a thorough understanding of even one ecosystem has never been achieved. Reaching this goal will involve every discipline of marine science.

Calypso *divers take a census of **a reef community** by injecting a harmless narcotic that renders fish motionless for a short period of time.*

The Ocean in a Test Tube

Chemical pollutants that enter the sea can become dispersed by currents but later reconcentrate in food chains, with harmful if not disastrous results. Case studies on reconcentration include one that measured stationary water masses in basins off southern California, revealing concentrations of lead in the upper layers (new water) as compared to the lower levels (old water). In spite of the advances that have been made in chemical oceanography since the *Challenger* Expedition took water samples, researchers are too often still compelled to conduct environmental chemical studies by collecting samples at the ends of mile-long cables hung over the sides of ships. The samplers attached to these cables have been made more elaborate; they take on various shapes and sizes and are made of various materials to protect the samples from chemical contamination. Researchers are now busy developing better instruments for automatic and simultaneous analyses of seawater. Simultaneous analysis would solve a

problem that has always plagued the chemical oceanographer: when samples are not analyzed immediately but are stored for future study, serious errors can be introduced.

A major effort to acquaint ourselves better with the chemistry of the ocean is in the Geochemical Ocean Sections Program (GEOSECS). By making detailed measurements of the properties of the ocean from the arctic to the antarctic at all depths, it will provide for the first time a set of physical and chemical data taken on the same kind of water samples. These data will provide material for analysis of ocean mixing and organic productivity and serve as a baseline for measurements of the levels of pollutants.

Intimately related to assessing the chemicals we dump into the sea is the study of currents which may confine or disperse them. In addition to the major currents of the sea, there are the small eddies that occur along coastlines, and there is the general circulation of deep-water waves. Many dumping sites and liquid-waste outfalls are based upon the concept that a passing current will carry away the pollutants and disperse them. In some cases these subtle circulation cells carry the pollutants around in the same general area and provide very little dispersion.

Currents have been studied for a long time, but the precision of the science is still very limited. In the past a common method of studying large ocean currents was to drop overboard bottles with a card inside directing the finder to note the time and location he encountered the bottle and send the information to the address indicated. Presently scientists base their estimates of water motion on such devices as propellers or other rotating mechanisms which record the number of turns and thus how fast the water was moving during a period of time. Although relatively accurate for particular locations, a

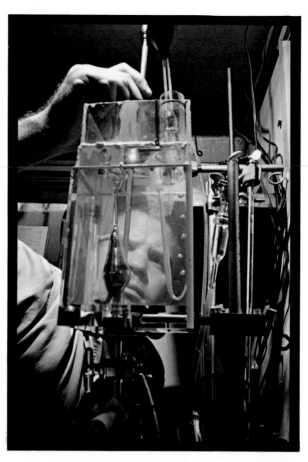

*A diver inspects a **current meter** suspended above the ocean floor (left) and a **chemical oceanographer** analyzes seawater for dissolved gases (above).*

great number of these devices must be used to get any idea of circulation patterns. As a result, scientists often depend on dye studies to learn how water moves in an area.

The circulation of deep-water masses is computed from salinity and temperature measurements, on the assumption that water would flow from where the water column was the heaviest to where it was lightest. To check these calculations, Dr. Swallow imagined a self-contained, small sonar buoy made of two materials—one less compressible, the other more compressible, than seawater: the buoy would float at a certain depth, like a cartesian diver, and ships at the surface could trace its migration with their own receivers.

101

Waste and Life

When limited quantities of biologically derived wastes, such as sewage, are dumped in the sea, they can be disposed of by natural processes. Such chemicals have existed for millions of years and life forms have adapted to their presence. Evolution has even produced methods of utilizing some of them as nutrients. Of course, when there is an excess of these wastes severe imbalances develop in the biological community and ecosystem. Those animals which can thrive on the wastes will survive and develop vast populations, while the sensitive ones will be choked out or overcome by the chemical pollutants. The healthy diversity of life is destroyed. It is consequently very important to study the sensitivity to released chemicals of all the animals in an ecosystem. This could enable scientists to predict the problem areas and possibly avoid them.

But the biologically degradable chemicals are not our major concern. What is most devastating to life in the sea are the chemicals synthesized by man and the heavy metals. These substances were not present during the long evolutionary process and as a result animals and plants are unable to minimize their effects or render them harmless chemicals; these poisons will remain in the sea for a long time, maybe forever.

For example, insecticides are specifically designed to affect the nervous systems of insects. Their marine relatives, the crustaceans,

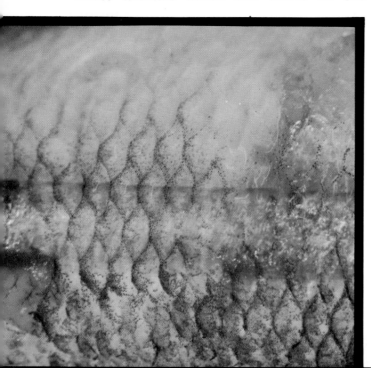

*The top smelt below suffers from an inflammation of the skin, a fungal infection, seen clearly in the closeup at left. Fungal infections increase as the resistance of fish to disease is lowered by **pollution**. In the photo at the top of the opposite page, a scientist studies the effects of various pollutants on sea life. A school of lookdown fish is seen opposite, below; their brilliant colors and healthy appearance indicate a pollution-free environment.*

are physiologically so similar that they too are killed, especially in their sensitive larval stages. Herbicides are now being washed to sea and in a similar fashion are killing marine plants. Many of these chemicals are extremely long-lived—some take over 20 years to degrade to half the original amount. As we continue to use them they will become more concentrated in marine life and sediments in coastal waters. We are now discovering that the chemical residuals of some are even more toxic than the parent compound. The impact of these pollutants can only be determined after studying each animal and its life history to see how and with what effects these chemicals enter their bodies.

A Record of Pollution

Not only is there evidence of man's impact on our shorelines and in dwindling populations of marine life, but a record of our pollutants is being kept on the bottom of the sea. Building up in the sediments are layers of samples of the effluents of civilization.

The process of natural weathering carries trace metals to the sea, where they are deposited. Extracting these metals from land has speeded by many times this natural weathering process. Through industrial and domestic sewage, excessive concentrations of toxic metals are being dumped into the sea. A recent study of the waters and sediments in southern California near sewage outfalls has shown that we are introducing, in some cases, as much as one thousand times the natural concentration of certain metals. Among these elements are zinc, copper, lead, cadmium, chromium, mercury, and even silver. The recent and extensive reduction in bottom-dwelling life forms near these sewage outfalls supports the view that marine life is being poisoned by these substances. Other measurements made in surface waters in mid-Atlantic, compared with similar ones made in 1925, prove that lead content in the high sea has increased almost five times! Through the chemical analysis of various sedimentary layers we are finding a historical record of man's impact on the sea.

As discussed previously, the best tools for obtaining geological samples of sediments are coring devices that return a column of mud undisturbed, allowing a positive analysis of each layer. To obtain a larger sample of the upper layer and the life in it, scientists use the bottom grab. One, called the orange-peel-bucket sampler, has jaws which resemble the segments of a peeled orange. Upon reaching the bottom, the jaws close, engulfing a volume of the bottom to be retrieved. Clam-

shell-type grabs are in more common use. As contact is made with the bottom, a powerful spring snaps the jaws shut, taking a bite out of the substrate.

To monitor the geological aspects of man's impact, we cannot limit ourselves to detecting trace chemicals. A growing problem is simply the presence of sediment loads on coastal waters. For example the rapid development around Honolulu, Hawaii, has laid bare to erosion the surrounding lands. The resultant sediments have a devastating effect on the delicate coral polyps, which smother because they are unable to cast off this material. The viable and active reef is converted to a deserted wasteland. The coral reefs of Florida are having similar problems.

Assessing the impact of man on the ocean environment is a major task that reaches into every field of marine science. No field of study is isolated from the others; to understand what we are doing to the sea requires a consideration of all its aspects.

*Oceanographers use **grabs** (top and opposite) to obtain large samples of bottom sediment. Analysis of these samples indicate that some of the pollutants we dump in the ocean are being concentrated in the upper sediment layers of the sea floor.*

Chapter X. Remote Sensing

Probably the most valuable technological advances in oceanography that have come in the recent years have been in the development of various remote sensors, by which we can study certain phenomena without needing a human presence on the spot. These remote sensors are rapidly becoming the eyes and ears of man. Data that they provide are fed into computers and then the data can be analyzed by scientists. The greatest advantage of remote sensing is in the fact that man is expensive and not expendable. Equipment is generally cheap and can be lost with no great consequence. A satellite that fails to achieve orbit and burns up in the atmosphere can be replaced. An astronaut that died in the accident would be irreplaceable.

The most spectacular remote sensors are the ones carried by satellites. Already they have been used to study currents and geological structures and to locate areas of high biological productivity in the oceans. Fishermen are able to obtain data from satellites on

---

"Oceanographic remote sensors are rapidly becoming the eyes and ears of man."

---

weather conditions, sea surface conditions, ocean circulation, the color of various ocean areas, and the location of fish, oil slicks, and iodine concentrations. The remote sensing instruments available now for use aboard satellites are basically modifications of aerial survey instruments.

Remote sensors are also carried by buoys at sea, which can hold instruments from surface to the deepest ocean for measuring, recording, and transmitting sea and weather conditions. A number of buoys are in use already, giving us an abundance of hitherto unavailable information at minimal expense. It seems probable that hundreds and maybe tens of thousands of additional buoys will be put to work in the next few years. They will be interrogated by passing satellites, and the satellites will later transfer all the collected data to one central computer.

Not only can remote sensors assist in the physical phenomena of the sea and observe its conditions from space, but one of them, underwater remote-controlled television, is becoming an important tool to marine behaviorists. Animal behavior is very difficult to study in the sea because the mere presence of a diver in the vicinity of the subject will cause most animals to cease most normal behavior patterns. It must be a terrifying sight for a fish to watch a large, clumsy, double-tail monster approach and then rest nearby, loudly spewing out bubbles. One solution to this, which has proved successful, is a stationary underwater TV setup which was used on Bimini Island. In a short time the fish become accustomed to the inanimate structure and could be observed in the lab ashore carrying out their normal activities. The installation was also equipped with an underwater speaker and recording system to allow the scientists to watch the reactions of fish as various sounds were played into the water. In addition to sharks—known to be very sensitive to sound —the device attracted groupers, snappers, and other predatory fish. Such a system allows scientists to record on videotape all responses to the sounds. In this case eliminating man from the sea has distinct advantages.

Remote sensors permit man to make observations of ocean phenomena that might otherwise be impossible. This dolphin is carrying a data transmitter.

Buoys

Acquisition of data by instruments dangled from a ship is an expensive procedure largely because of the cost in ship time. If the instruments are suspended from buoys moored at the surface of the sea, data collection is cheaper and much long-term data can be acquired from an array of them.

Buoys, or buoy systems, can be placed at the ocean surface or on the bottom. The buoys store information obtained from sensors attached to them. Sensors attached to surface buoys record weather conditions, surface currents, and waves among other things. Subsurface buoys permit accurate measurements of currents, temperature and salinity, sound velocity, and so forth.

Information obtained by sensors on buoys can be collected and stored within the system until it is retrieved by a surface vessel. Newer systems transmit data to stations onshore or to satellites.

*Small **bumblebee buoys** like the one at right have helped oceanographers gather simultaneous data from several areas in the ocean at a reasonable cost.*

*A **monster buoy** (below) collects a wealth of data. It can function in winds up to 160 mph and in seas 60 feet high.*

Platforms attached to the sea floor are also being used for various oceanographic measurements. These platforms are relatively cheap in comparison with ships and they are stable, permitting long-term, uninterrupted measurements.

A monster buoy might have sensors for making as many as a hundred different measurements in the air and water.

The Naval Oceanographic-Meteorological Automatic Device (NOMAD) can remain

unmanned on station for months at a time, relaying data to shore facilities. The 11-ton, aluminum-hulled buoy is equipped with a complex of electronic instruments in its floating framework.

Buoys will probably play an increasingly important role in ocean research. There is too much variability in the sea to generalize from a few isolated spots about what is happening on a global scale. We need a broad picture of the grand scheme of things that could easily be obtained by saturating the oceans with aircraft-dropped inflatable buoys which could send out basic oceanographic data. The centralized flow of information would enable scientists to follow minute by minute the genesis of global weather systems.

Colorful problems arise when instruments are left unguarded. They become prime objects for scavengers, or they are mistaken for food by predators, or they are considered valuable salvage or dangerous weapons by humans who come across them at sea.

Robots on Leashes

In 1966 a tethered unmanned submersible called CURV helped in the search off Palomares, Spain, for a hydrogen bomb lost at sea after a refueling accident involving two U.S. Air Force planes. It was CURV that managed to grab the bomb with its mechanical arm and bring it to the surface. This was just one of the more spectacular jobs that this kind of submersible has been doing lately. CURV stands for Cable Controlled Underwater Recovery Vehicle. Its normal mission is to recover spent torpedoes and missiles. CURV sees with television and sonar, moves with propellors, and wields a mechanical arm that it can shed if it must.

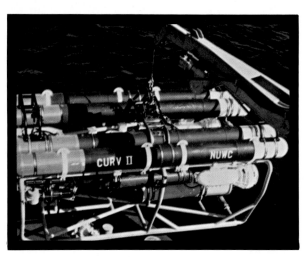

A similar French-made instrument, the Telenaute, was used by the *Calypso* team, to supplement their third exploration of the Fountain of Vaucluse, a deep siphon sloping down in the earth's entrails. The *Calypso* divers stopped their investigations at a depth of 300 feet, and from the surface they remote-controlled the delicate descent of Telenaute through narrow corridors, down to a depth of 420 feet.

Gloria, a submersible belonging to the British National Institute of Oceanography, is a tethered unmanned submersible that functions as a seeing ear. Gloria can take accurate pictures of the deep-ocean bottom over a strip 12 miles wide. The submersible is towed at 10 knots at the end of a 600-foot cable. Gloria's steering system enables her to stay on an absolutely even keel.

Mobile robots are another form of tethered submersible. The "Mobot" has two arms for grasping and two for turning wrenches.

A six-limbed sea robot has been proposed that would operate from the end of a cable leash. Two of its limbs would serve as manipulators and the other four legs could permit the robot to stand up. In addition to the normal functions of an oceangoing robot, this one would also be able to take core samples.

There should be a great industrial future for the tethered unmanned submersible. There are advantages to keeping man on the surface in some cases and letting an instrument do the underwater work. The range of depth of the diver has been dramatically extended recently, but beyond 2000 feet a man cannot be used directly. In a submersible he must sit behind the hull and operate some sort of manipulator, and this is not easy when the submersible and the manipulator are moving simultaneously. This arsenal of robots, CURVs, tethered unmanned submersibles, is really designed to perform programmed industrial tasks, such as those that face the petroleum industry today. But the unmanned submersible can never replace the manned vessel since no machine can ever compete with man himself for accurate evaluation and quick decisions.

CURVs, tethered unmanned submersibles like the ones shown here, have as their normal mission the recovery of spent torpedoes and missiles, but a CURV has also retrieved an H-bomb lost at sea.

Photographs from Space

Of all our devices for remote sensing, the photographic equipment developed to give us pictures from satellites orbiting the earth will ultimately prove most valuable.

Color photographs from space have been taken with various filters in order to help display a specific phenomenon. Today full color pictures are analyzed in the laboratory and are treated there by computers to produce several pictures from the same shot, each showing a different aspect of the same area. They have already revealed where sediments and pollution discharged by rivers drift into the sea. They show the daily meandering of the Gulf Stream, its beautiful blue being distinguishable from the neighboring, more productive green waters. This ability to record, enhance, and analyze water color from space may help us to follow the displacement of the fishing grounds and constantly reassess the maximum yield of the sea compatible with long-term global management policies that will be developed in the not too distant future.

Satellite photographs have already been used in surveying shallow water areas, properly locating shoals and reefs, studying geological features, measuring heat flow from the sea surface, and interpreting surface detail to reveal the local effect of wind and tide and the transportation of sediments. Photographic chromatography, as it has been developed recently, extending broadly in both nonvisible ends of the spectrum—infrareds and ultraviolets—is the technique that has enabled us to derive so much information from satellite photographs of the earth.

*This **infrared photograph** of San Francisco Bay reveals more information than would an ordinary color photograph of the same area.*

Looking Back

Not long ago a disastrous drought hit parts of Africa. From a satellite scanning the earth, chromatographic analysis of "photographs" revealed the possibility that there were hitherto unknown sources of underground water in the area. It was too late to relieve the immediate situation, but it was hoped that determination of the existence of the water sources would assure that when drought occurred again an answer would be ready.

The satellite that made the discovery was ERTS—the Earth Resources Technology Satellite. Of the countless discoveries that ERTS has made since it was launched in July 1972 this is one of the more spectacular. The satellite is outlining a census of the natural resources man needs for survival.

Each day the one-ton satellite makes about 14 revolutions around the earth. Coverage of the whole globe occurs every 18 days as it travels in an orbit near to the poles and synchronous with the sun. (As the earth rotates beneath the satellite, the coverage proceeds westward.) ERTS orbits 570 miles above the earth, and the images provided by its remote sensors over a 3200-square-mile site in Arizona were able to discriminate 29 separate types of vegetation. Elsewhere, investigators found that the images sent back could distinguish between fields that were fallow, freshly plowed, newly sown (detectable by recent irrigation), or bearing crops.

ERTS is providing valuable information about ocean currents, sediment distribution near shoreline areas, and large upwelling areas. These will be mapped and monitored to determine their relationship to fisheries. ERTS is also monitoring pollution of coastal areas of the oceans as well as inland seas and waterways. Parts of the world that have

Different wavelengths reveal different characteristics of sea and land in these photographs of Rhode Island and vicinity taken from the ERTS satellite.

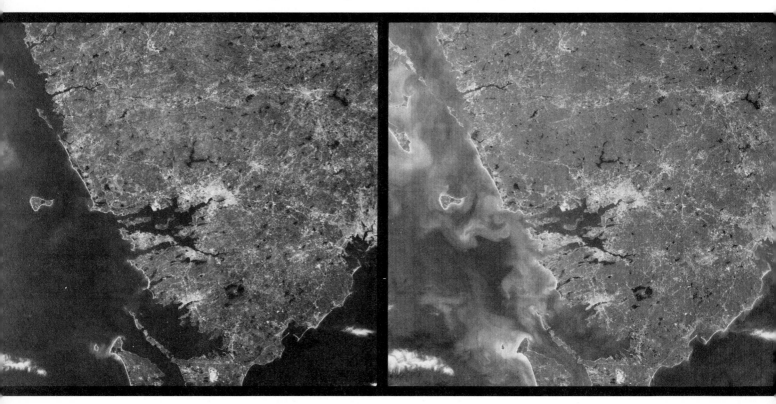

*A black-and-white negative of Osaka, Japan, **taken from the ERTS satellite,** is made into a color composite, revealing unapparent characteristics.*

never been mapped are being accurately described for the first time. Ocean dumping and surface pollutant films are being plotted around New York, including acid-iron wastes, sewage sludge, and suspended solids. Other images showed a 15-mile patch of iron oxides and sulfuric acid dumped at sea off Corsica. Near New Jersey ERTS discovered tide-washed sewage threatening beaches. It has monitored ice floes in the Arctic and detected plankton in the Atlantic.

A major objective of the program is to outline the first comprehensive inventory of the earth for future reference. Because the high-altitude photographs are difficult to correlate to a definite condition in the sea, or in a forest, scientists have located themselves on the ground at the exact spot where the satellite is taking pictures. The scientists too are pho-

tographing and making detailed analyses of that geographical point. They are collecting "ground truth data" which can be used in the future to correlate similar space pictures to actual conditions on the ground or sea. Our recent expedition to the antarctic permitted us to participate in this important project. The *Calypso* and its crew were able to collect some oceanographic data in the antarctic when our course crossed the path of the satellite. Among other things we measured the chlorophyll content of the summer antarctic waters and then beamed our information to a satellite above, which relayed it to NASA headquarters in the United States.

115

Partners in Research

Recently, a yearling gray whale was given its freedom after having spent most of its life in a Southern California oceanarium. Her name was Gigi and she was to join her kin as they migrated north to the Bering Sea. Her departure was heralded by men on ships and in planes waiting and listening for signals she was to send. Gigi was equipped with an instrument pack that contained devices to measure temperature in relation to water depth and a transmitter to relay the data to the scientists. This system would also allow them to monitor Gigi's swimming path. Unfortunately problems arose with the transmitting equipment and Gigi's swimming behavior was erratic, limiting the number and strength of transmissions.

The large gray whales often remain submerged for 25 minutes or more, while most

A gray whale, equipped with data acquisition and transmission instruments, lies in its protective cradle just prior to release.

porpoises surface from every 30 seconds to 6 minutes, and frequently several times in succession before diving again. Pilot whales rarely dive for more than 8 to 10 minutes and remain on the surface for long periods of time before diving again. Theoretically, with porpoises and smaller whales fitted with instrument packs, then, transmission is frequent, and they are less likely to move out of receiver range and be lost during a dive. Some specialists concluded that small-toothed whales would be the best subjects as remote instrumentation platforms. In addition to their short diving times, these whales feed on fish which are of greater interest to fisheries biologists searching for productive fishing grounds than are the krill sought by gray whales. The toothed whales could still provide a sufficient sample of the water column since they dive to considerable depths. The dolphin has been used in subsequent tests in the hope that it would be very valuable, particularly in locating areas of high biological productivity. Observations of free dolphins also proved that often there is a pattern between the dolphin's diving behavior and the daily vertical migrations of creatures indicated by the famous "deep scattering layers." Unfortunately, the dolphin is gregarious. Even a miniaturized instrument package attached to its back handicaps the animal dramatically. The few minutes taken to capture the dolphin and release it with the transmitter are enough to isolate it from the pack. If it rejoins the pack, it is not always accepted by its peers, and the instrument and antenna sticking out turn the animal into a clumsy clown. The dolphin's stays at the surface for air are also very short. To be instrumented efficiently without interference with behavior, a marine mammal has to be large enough to carry the miniaturized instrument package without any notice whatsoever. The pilot whale is certainly a good choice. The orca is too rare and too clever; its family would tear the instrument to pieces. The sperm whale is ideal, because it dives very deep and stays for a long time at the surface.

The day may come when flotillas of small buoys or the monster stations will be obsolete. Instead of being replaced by machines they may be supplanted by large marine mammals cruising the sea at will, silently transmitting information about the environment. The marine mammal could be man's partner in a quest for knowledge.

*The **transmitter pack** being fitted to this **gray whale** will send back information on water temperature as well as a record of the whale's route.*

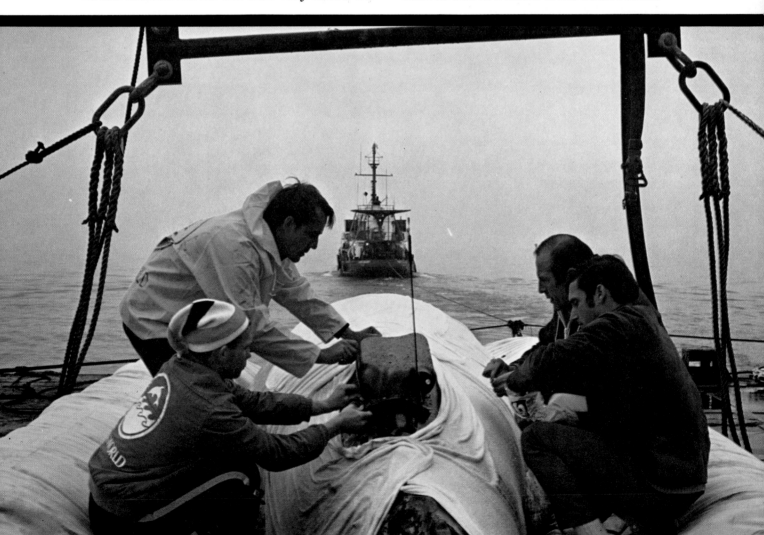

Deep-Sea Photography

Before the deep sea was accessible to small research submarines, our only glimpses of the bottom and life in the abyssal depths came from photographs taken by cameras lowered from above. As a substitute for direct observation, cameras are extremely useful and much less expensive: they are the bathyscaphes of the "poor" oceanographer. Dr. Harold Edgerton of MIT, the famed pioneer in strobe lights, has been the most imaginative and obstinate mind in turning deep-sea photography into a practical, indispensable tool.

Oceanographers today generally use cameras that can take hundreds of pairs of color stereo pictures at considerable depth before being brought back to the surface. Powerful strobe lights are attached to the cameras, and the whole package is connected to a cable. In some, perfect focus is assured by a sonar system that sends out sound waves, or "pings." Each signal travels both downward and upward. The first one heads for the bottom, strikes it, and then bounces back to the ship. The upward signal goes directly to the ship. Knowing the speed of sound in water, the cameraman measures the difference in time between the two pings re-

ceived and can tell how far—within one foot—the camera is from the bottom. The Edgerton deep camera has also been used on board *Calypso* in two other dramatic ways. As early as 1959, we attached it to a special sledge and towed it along the Mid-Atlantic Rift Valley for three miles in 10,000 feet of depth, obtaining hundreds of pairs of color stereo pictures. Later we used the cameras attached to a deep-midwater sled, towed at five or six knots for three hours across the deep scattering layers.

One of the areas where undersea photography has been of special value is archaeology. Photographs provide a permanent record of what has been found and its exact location in relation to other objects.

Television cameras have also been adapted for use in the ocean, but they have severe limitations at great depths. They must be connected to the surface by special, fragile coaxial cables, and at significant depths they require power relays along the cable. Also, television signals that are easy to transmit with radio frequencies in the air cannot be transmitted underwater without cables because the sonar frequencies are too low. But TV cameras can be set on the bottom in shallow water. Here innumerable important observations can be made with the opportunity of immediate analysis. Underwater television is especially useful in situations where men topside need immediate feedback on what is happening below. For example, in monitoring the activities of divers, inspecting a pipeline, or observing marine organisms, the scientists or engineers depend upon their artificial eyes—the camera.

New television tubes that amplify light 30,000 times make it possible to eliminate artificial lighting, thereby eliminating the backscatter of light by small particles in the water which reduces image resolution.

The **"troika"** (top) is ready to be lowered from the Calypso. *It is a large, stable sled, equipped with an Edgerton automatic-flash camera. It is asymmetric in order to keep the towing cable out of its field.*

H. E. Edgerton and Captain Cousteau *adjust a strobe-light reflector* (above) *for a camera mounted on a bottom sled, the "troika."*

Snoopy (right), *a remote-controlled underwater television system, is one of many such designs that allow scientists to get a view of bottom configuration and of animal population and behavior.*

The **television camera** *can now replace man in much exploratory work, such as searching for wrecked ships or making detailed studies of the ocean floor* (left).

Chapter XI. No Substitute For Man

Man cannot leave the study of the sea to automated mechanical or electronic equipment. Dials, computer readouts, and cathode-ray tubes can tell the scientist much of what conditions lie beyond his range, but there can be no substitute for man's occasional presence in the sea. Personal experiences often lead to intuitive insights that may solve perplexing questions or open up totally new fields of thought about the sea.

In addition, man in the sea can detect subtleties and make quick decisions that no technology will ever surpass. Laboratory observations are limited by experimental design; an accurate duplication of the ocean environment is difficult to achieve and presupposes an understanding of all factors influencing the subject of study. Hypotheses resulting from controlled experimentation

---

"Always the most important factor is man with his curiosity about the ocean world."

---

must eventually be tested in the actual environment. Therefore, it becomes essential to put the oceanographer in the sea to test his hypotheses and develop new ones.

The scientific activities of man beneath the sea might be said to have begun in 1934 with the descent of Beebe's bathysphere. The next most significant step was taken in the 1940s with the development of the fully automatic regulator valve that permitted divers to carry their own air supply with them. The Aqualung untethered the diver, giving him increased mobility and range. For more than 20 years since then scientists have been able to apply direct underwater observation to a wide range of research projects, increasing both the quality and quantity of informa-

tion on the marine environment. Nevertheless, the application of *surface-based* diving to science has its limitations in depth and duration by the requirements of decompression. Furthermore, surface-based operations are sometimes restricted by weather and sea conditions. Finally, not all scientists are physically or emotionally suited for diving.

To reduce the limitations imposed by depth and the time consuming processes of decompression, our efforts have taken us in two directions. The first, underwater craft, are continually being built. There are now more than 150 submersibles designed especially for ocean research. The second, underwater habitats, are the latest addition to the arsenal of direct observation tools. They provide the scientist with the necessary refuge for rest and resupply, while affording unlimited access to the environment. Living and working underwater, the scientist is able to make observations over extended periods of time and to undertake sophisticated experiments that could not be carried out in a laboratory. When Tektite II had completed its seven month effort in 1970, over 50 scientists had each spent up to three weeks carrying out detailed investigations of many facets of tropical oceanography.

The free diver, the research submarine, and the habitat are the best means we have of exploring the oceans in quest of inspiration for new research departures. But none of these systems is an end in itself. Always the most important factor is man with his curiosity about the ocean world.

*Man's ability to perceive, decipher, and understand has made him a **data acquisition system** for which there can never be a technological substitute.*

Exploring the DSL with *Deepstar*

Of the many submersibles in operation for science, one that has made a substantial contribution to our understanding of the ocean is *Deepstar*.

Deepstar has done some very valuable research on the deep scattering layer. In 1965,

using the diving saucer, Dr. Eric Barham, one of the world's leading authorities on the identity and behavior of the tiny marine organisms that compose the DSL, had made observations of horizontal and vertical movements of plankton and fish and their daily

to two minutes for a "light look." Two-minute dark periods were interspersed. In this way the scientists were able to see and to photograph the animals and also to move down slowly during the dark periods and in this way not influence the movement of the animals attracted to the light. After making several dives during the day and at night, Dr. Barham found that in general the upper layer of the two-layer system in the San Diego Trough, 30 miles at sea, showed a majority of hatchetfish, while the lower layer comprised mainly siphonophore jellyfish. It appeared that sound was scattered by both animals, but Dr. Barham believes that the primary scatterers were the fish.

It was easy to maneuver *Deepstar* because it has good depth control and it is able to remain almost motionless in the water column. Thus the scientists could watch the animals from only a few feet away without disturbing them. On the third dive of this series, *Deepstar* and Dr. Barham hung at 400 feet for over an hour, turning lights on and off and photographing the animals as they moved about in their natural environment.

Deepstar also explored gullies, canyons, and steep continental slopes where geologists needed to go slowly down the bottom to search for marine terraces that had been indicated on echo-sounder traces made from the surface. By identifying the terraces and their exact depths, geologists could reconstruct the histories of the continents and acquire information on how sea level has varied on a worldwide basis.

migrations from a depth of 350 fathoms to the surface. The technique that Dr. Barham established was first to survey the area with an echo sounder from a surface ship in order to determine the position of the sound-scattering organisms. A series of transects were then made in which the *Deepstar's* lights were turned on every three minutes for one

Deepstar 4000, *with its special acoustical sound array, has done valuable work on the deep scattering layer, in gullies and canyons, and on steep continental shelves. The propellor-driven submersible carries a crew of three and can cruise from 6 to 12 hours at a speed of three knots. Deepstar has good depth control and is easy to maneuver.*

Tools of the Archaeologist

Ships have sunk every year since man first floated down the river on a raft. Entire cities have vanished beneath the waves. The possibilities of undersea archaeology are enormous, and some remarkable discoveries have been made as new tools have been developed. Sonar, both side-scanning and sediment-penetrating, underwater television, core samplers, magnetometers, metal detectors, and submersibles—all have proved valuable, but none so much as the relatively simple development of free-diving itself, allowing the archaeological worker unencumbered access to the object of investigation.

The first step is obviously to locate sites. A crude method in use since the turn of the century—helmet divers hanging under a boat and towed slowly through the water—

*The **airlift,** which acts like a giant vacuum cleaner, is a valuable tool for archaeology underwater.*

has been successful in locating the famous antique wrecks at Antikythera and in Mahdia. Later, our aqualung divers also used the towing technique in Corsica and Tunisia. Then towed submersibles were developed, such as our wet device or dry ones like the *Towvane*. They are connected in both cases to the ship by a cable or sonar telephone. Probably a more efficient method of searching for sites is with a combination of side-scanning sonar and closed-circuit television. The sonar cannot always distinguish between wrecks and outcroppings of rocks. Television, guided by the findings of the sonar to a possible site, can. Acoustic imaging devices are now being developed to make "photographs" with sound waves.

A land archaeologist needs a shovel, but for the worker underwater it is the air lift, first used in our 1953-54 excavation of the Greek wreck at Grand Congloué Island near Marseilles. Since then it has been used on virtually every major undersea excavation. It is a suction hose, a vertical pipe or tube that acts like a vacuum cleaner. As air pumped through a hose from a compressor on the surface enters the tube near the bottom, it rises toward the surface and expands as the pressure decreases, causing suction at the mouth of the tube. This suction pulls in water, as well as sediments and other material small enough to enter the tube, and clears the wreck site. Submarines, already used to map the visible portions of sites with stereophotography, will soon be used to clear away sediments with portable, neutrally buoyant air lifts directed from the submarine by remote control.

Our work on the Grand Congloué wreck demonstrated that it was physically possible for free divers, if the time they spent underwater was strictly controlled, to work thousands of man-hours with little or no danger of decompression illnesses.

A diver recovers two small black Campanian pottery bowls from the wreck of an ancient Greek sailing ship sunk at Grand Congloué.

But probably most important for the future of underwater archaeology is the use of saturation diving and undersea habitats. The day is not far off when archaeologists will live for weeks at a time beneath the sea. Or, where underwater stations are not practicable, divers will be lowered in pressurized personnel transfer capsules from which they will enter the water to work. At the end of their work day they will return to a shore- or ship-based pressure chamber to eat and sleep. In this way it will be necessary for them to decompress only after several days at the least. Mixed-gas diving, substituting helium for the nitrogen in the air supply, allows for deeper dives than ever before and reduces the likelihood of narcosis.

Whatever technological advances are made, archaeological digging operations are not aimed so much at bringing back objects from the bottom as at carefully investigating the relative position of each component to allow a reconstruction of the site or the ship. Reconstructions can lead us to fascinating discoveries about past cultures.

Fishing nets (above) *could be replaced by traps, which are efficient and easy to handle.*

When there is no bait in the trap, **curiosity alone** *will entice a fish* (opposite top).

Bait Is Not Important

William High has as part of his responsibilities the well-being of the sablefish industry in the northwest United States. He was part of the Tektite II mission in Lameshur Bay of the Virgin Islands, and his job was to discover what he could about the behavior of reef fish in relation to traps and other fishing gear. He made some remarkable discoveries.

Most astounding, probably, was that bait in a trap meant little or nothing to a fish. For centuries the fishermen of the Virgin Islands, like fishermen everywhere, have been baiting their traps. High's observations showed that the fish came in out of curiosity or out of apparent desire to occupy a new space, whether or not there was bait in the trap.

High also noticed that a fish caught in a trap seemed to act as a visual stimulus to other

fish to come in. Eventually, when a critical point of crowding was reached (the experiment was done on squirrelfish), they began to make frantic movements in a rapid darting fashion, which High believes was to warn others to stay away.

It was also discovered that small fish caught in a trap tend to attract larger ones. Thus it might be possible somehow to lure small fish into a trap and provide them an inner sanctum in which they would continue to attract larger fish and yet not be eaten by them. It was found that the catch rate was dependent primarily on positioning of the trap. If traps were placed a few feet away from the traffic patterns of reef fish, none of the fish would be caught. Hence the value of the diver in this situation—he can be sure that the trap is in precisely the most advantageous position to catch fish.

*A biologist counts fish in **an experimental trap** (below). Small fish in the trap lured larger ones.*

Sometimes traps in the experiment were placed in what was apparently the domain of a fish—often a grouper—and the fish would either go inside it and take up a defensive position or else it would remain outside the trap and adopt highly aggressive behavior toward the fish that entered it.

Some of the observations made by the scientists of the Tektite II mission were made while sitting inside the habitat and letting television do the work. They soon found, however, that television can do nowhere near what the diver can do. The diver was able to track the fish over long distances —something the television could not do. Furthermore, television has night blindness. A diver can see far better in the dark. Most important, however, are the myriad of incidental observations that the man on the spot can make and that television ignores.

A Close Watch on the Grazers

It was morning and the plant-eaters were stirring. Gray angelfish were among the early risers. The diver paid little attention to them but was intent on the activity of three large triggerfish that had just swum into a wiremesh cage. The cage covered a half-dozen species of plants and was designed to exclude grazers like the triggerfish.

The diver was a member of Tektite II, the research program operating from an underwater habitat in Lameshur Bay, the Virgin Islands. Her experiments gave us further evidence that the diver was absolutely essential for precise study of many ocean phenomena. Dr. Sylvia Earle Mead, a diving scientist, wanted to discover the relationship between benthic plants and herbivorous fish on and around a coral reef. No amount of surface sampling could have produced the same results as Dr. Mead's presence. Accurate information about the arrangement of the plants and the activities of the reef fish required an on-the-spot observer.

Dr. Mead averaged six hours a day in the water and some days spent as much as ten hours working on the reef. Day and night observations were made to watch fish over a 24-hour period. Dr. Mead found that plant grazers were active only by daylight; at night the herbivores rested in crevices and hollows in the reef. She observed what plants they ate and how the distribution of plants affected the distribution of fish and vice versa. The feeding behavior of various fishes and their choice of diet were studied. Samples brought from far away were offered to the fish. Some were accepted and some were rejected. Definitive preferences were exhibited. Habits of thirty-five species of herbivorous fish in 14 families were recorded.

Wire cages were put on the sandy bottom to protect the plants from grazers; more luxuriant growth ensued within the cages than in the adjacent area. It was concluded that the selective grazing of fish influences the abundance, distribution, and diversity of reef vegetation. Herbivorous fish are abundant on the reef, where there is cover and protection from large predators. Few venture far from the reef's safety.

The abundance and diversity of plant life increases *away* from the reef. There were 154 species of plants counted in Lameshur Bay and about 30 on the reefs. Immediately surrounding the reef there is an area almost devoid of plants, apparently the result of heavy grazing. In areas where physical requirements for plant growth are satisfied, at-

tached vegetation is abundant where there are few grazing fish, and where grazing fish are abundant, vegetation is sparse. It was further discovered that the role of invertebrate grazers on plants is small relative to the impact of grazing fishes on the reef. (In cold water the situation is reversed. Off the California coast sea urchins and molluscs eat more plants than the fish.)

*Diving scientists of the Tektite II program have placed cages over the sandy bottom to study the effects of **grazing on algal growth.** Exclusion of grazers permitted luxuriant algal growth. Notice that there is more vegetation around the cage marked 300 ft., which is further from the reef (above left) than the one marked 600 ft. (above right). This occurs because grazers seeking the reef for refuge graze more on nearby areas.*

*The lack of obvious vegetation on **a typical reef** (below) is indicative of reef grazer activity.*

What the Blue Holes Tell Us

Some of the ocean's most curious phenomena are the blue holes, probably formed by the collapse of limestone caverns. Divers have investigated the blue holes and in doing so have made some notable contributions to geological oceanography.

The term "blue hole" comes from the relatively blue color of the water produced by the depth of the holes as compared to the green shades of the surrounding shallows. During a rising tide, water about 18° F. cooler than the shallower water around them upwells from the holes. During a falling tide, the water is sucked into the holes slightly.

The blue holes were probably formed during the ice ages of the Pleistocene epoch, when the sea level was much lower than it is now. As great glacial ice sheets spread over the continents, so much water was taken from the oceans to form them that the sea level fell. The holes, or caves, were carved out when fresh water trickled through the limestone sediments; as it dripped, it dissolved some of the calcium carbonate, which formed stalactites and stalagmites. In some areas, particularly the Bahamas, great networks honeycomb the limestone below ground level. Seawater fills these caves, and where the ceiling has collapsed, abundant sea life lives in the exposed "blue holes." The tidal flux in these blue holes is puzzling; some have great surges of water during tidal changes while others remain relatively calm. Divers from *Calypso*, investigating the Lighthouse Reef blue hole near British Honduras, have found stalactites that are inclined 15°, giving an indication of plate-tipping and continental drift. It can be assumed that these stalactites were originally in a vertical position and that there has been a tipping of the region about the same number of degrees to the south.

A diver from Calypso *examines a stalactite hanging in a **blue hole** above. Blue holes have yielded many clues about the past history of the earth.*

Studies of the Lighthouse Reef blue hole have also made it possible to record fluctuations in sea levels between the present level and one that may have been 500 or 600 feet below the present stand about 18,000 years ago. The stalactites, calcium carbonate columns, and drip-stone encrusted walls that divers have discovered here could only have been formed when the structure was above sea level. It is supposed that the blue hole was originally a large cavern similar to those on land. In time, as the hole became larger, the roof was no longer able to support it and collapsed. When the glaciers melted in the end of the Ice Age, sea levels rose and the entire structure became submerged.

The blue holes not only can provide us with important scientific information, but they are also among the greatest spectacles that we have encountered in the underwater world. Efforts are underway to have some of them protected and declared part of an underwater national park.

*A diver tries to measure how much this **stalactite** deviates from a vertical plane (below). Some tilt as much as 15°, giving an indication of continental drift.*

Aquanauts *carefully escort a stalactite to the surface (above). These columns of calcium carbonate were formed during the ice ages.*

Observation Bubbles

An unusual experimental vessel doing oceanographic observations today is the U.S. Navy's *Sea-See*, a catamaran-type craft with a spherical underwater observation compartment. The vessel provides an excellent platform for relaxed, near-surface investigation of the behavior of marine mammals and sharks, two areas of considerable interest.

The navy has for a number of years carried out research on marine mammals, for many reasons. The dolphin, for instance, is an animal that has abilities far beyond our own in underwater communication and acoustics, which it uses in ways we understand only minimally. Observing the dolphin in its own environment could guide us to new concepts in electronics and mechanical equipment that would improve present navigation and detection systems and allow us to become more at home under the sea.

*The **Sea-See** (top) is a catamaran-type boat with an underwater observation compartment.*

***Protecting men from sharks** (above) has been a prime concern of the Navy's research programs.*

Another reason for the navy's cetacean studies is that these animals have a remarkable ability to detect, locate, and identify targets and to travel great distances to reach them: cetaceans have capabilities and mechanisms for distant and accurate underwater identification of objects vastly superior to anything yet devised by man.

The navy has been quite active in seeking a repellent to protect the downed flier or the shipwrecked sailor from sharks, for there have been many gruesome accounts of men being torn apart by ravaging sharks. One repellent that had been developed was found to discourage lemon sharks but to attract tiger sharks. The navy has a purple dye called the "shark chaser." When spread in the water, it is supposed to confuse sharks and keep them away, but this repellent is known to be practically ineffective. We simply do not know enough about shark behavior to make effective deterrents.

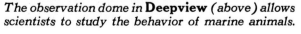
*The observation dome in **Deepview** (above) allows scientists to study the behavior of marine animals.*

Nemo, *top and bottom right, can be operated as a tethered or untethered submersible.*

Nekton Blasts for Oil

The two-man submersible *Nekton* recently went oil prospecting in Glover's and the Barrier reefs off the coast of British Honduras. Samplings from these areas have proved positive—oil must be there.

Nekton's scientists spent 120 hours underwater to detail the profile of the reefs and measure the depth of reef-building organisms. They brought up for carbon-14 dat-

ing more than a thousand pounds of rock samples blasted from eleven sites.

The samples proved relatively youthful—aged from 5000 to 10,000 years. They all came from reefs that at one time had been living coral. As the seas rose at the end of the last ice age, the corals died. Their stony skeletons were smashed by storms and tumbled down the reef front where they were impacted with calcium from marine plants and other sediments to form a solid limestone.

Nekton, nestled securely on the stern of her mother-ship Dawn Star *(above), helped uncover important information on coral reef formations that will aid geologists in their search for oil.*

Submersibles have increased man's ability to study the soil of the ocean floor and marine life. Nekton (left), basically simple, is rugged and is generally more reliable than more sophisticated machines.

The seaward reefs hardened to nonporous stone rapidly, while their counterparts in lagoons remained porous.

This exploration of living reefs has given oil geologists a better understanding of the ancient reefs now found on land in such places as the Swiss Alps and Texas. Today about half the petroleum found on land is located in porous limestones. Drilling is generally abandoned when nonporous rock is struck. The findings of the *Nekton* scientists may lead to the resumption of oil prospecting in locations that were previously thought to be worthless.

The Caribbean reef yielded yet another surprise to the investigating team. In cavities in the seaward slope and reef cliff they found an unusual formation of crystal. The calcium carbonate crystals are apparently due to the saturation of the surrounding waters and are deposited in crevices in a continuing process. The crystal packing was observed in water deeper than 250 feet, below the level at which photosynthesis takes place and reef corals flourish.

Prior to the *Nekton* studies crystal-filled cavities in ancient reefs could not be explained, and geologists believed that the reefs had hardened into limestone long after they had been thrust into the air and then buried by terrestrial debris. Now we know that the hardening takes place continuously in the seaward reefs.

Chapter XII. For the Sake of Knowledge

Man as a species has progressed to this point only because of his ability to keep written records. The wheel does not have to be reinvented every few generations. A young scientist can rely on the work of the past, on basic principles that need not be proved again; he can pick up where his predecessors left off. Today science proceeds at a rate limited only by the number of men and computers that can be put to work.

As basic research advances, applied scientists are a few steps behind, trying to do the job of converting the data supplied by pure researchers into practical applications for the advance of civilization. Keeping the pool of

---

"Keeping the pool of knowledge filled enables us to face the challenges of the future."

---

knowledge filled is our only security against the challenges of the future.

Throughout history there have been skeptics who questioned the value of scientists working in isolated laboratories on obscure projects. But the fantastic advances in science which only a few years ago were science fiction point out the inability of anyone to predict the future of scientific discovery. We cannot be so naive as to decide today which field of science is the most deserving of support. Research conducted purely for the sake of knowledge provides a storehouse of information. It is our duty to future generations of mankind to continue to faithfully nurture and replenish this storehouse of knowledge, never letting it run dry.

More and more, pure research is contributing to the practical aspects of scientific development. Only by experimenting with extremely cold liquid gases have we found that as some substances approach absolute zero they offer no resistance to an electrical current—they become superconductors. When applied to its fullest, this principle may revolutionize the field of power transmission and electronics, on which we are increasingly dependent. Experimenting with quantum mechanics and substances that can emit light led to the development of lasers, which are becoming more and more useful daily. With them we can project a three-dimensional representation of an object so that the viewer can see it in the round. Lasers can also measure the distance of the moon from the earth exactly. Studies on the nervous system of a large nudibranch are providing us with insights into the basic functioning of brain cells. Routine measurements of water conditions in the Red Sea led to the discovery of one of the world's greatest concentrations of minerals. Biochemists working on obscure marine animals are daily finding sources for potentially miraculous drugs. These are examples of pure research that have found practical applications.

Understanding the spaceship-planet we live on should be as much a goal of mankind as living in space and traveling to distant stars. And the oceans are the bulk of the living part of our earth. There are no greater mystery tales than the accounts of scientists in pursuit of elusive explanations for what occurs in the world around us. Moments of intellectual insight that result in exciting discoveries are to be cherished.

Studying the ocean has already helped man unlock some of life's secrets. Knowledge gathered for the sake of knowledge is in fact man's greatest practical asset; pure research is the key to the future.

Plate Tectonics

The movement of the earth's crust is of extreme scientific interest today. But one may question the direct importance to society of such a field of research. How is our life going to be improved by knowing whether the Americas were once part of the same landmass as Europe and Africa or if Australia was once a part of Asia? In the beginning of continental drift research its proponents were concerned little with such utilitarian questions; their work proceeded purely for the sake of science. But as with almost every new field of endeavor the technology developed or the information obtained proved to be of use to society at a later date. So it was with continental drift—now called plate tectonics. By gaining an understanding of how new material wells up through the crust to the ocean floor, scientists have gained insights to the formation of metal-rich ores and

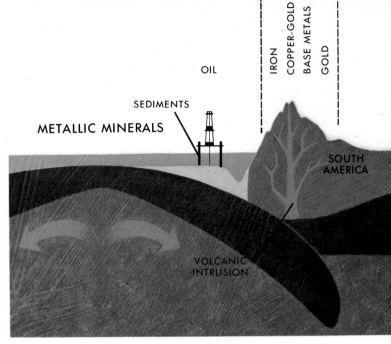

may be able to predict and locate metallic deposits more easily. By drilling in the deep sea, scientists have found that oil is not limited to shallow sediments next to coastlines. The techniques of drilling will expand research efforts to continue toward even more discoveries inconceivable at this time.

138

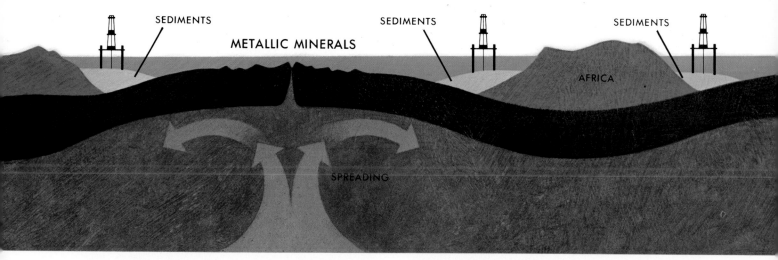

SEDIMENTS

SEDIMENTS

SEDIMENTS

METALLIC MINERALS

AFRICA

SPREADING

None of these economically advantageous results were intended when research began on movements of continents nor could they have ever been guessed. Only by probing the unknown in the interest of knowledge can all realms of science complement each other and begin to make a better life for us.

The illustration above represents limited evidence of a hypothesis indicating that the location of minerals in the crust can be predicted from the boundaries and activities of **plate tectonics.**

Fossil-bearing rocks *(left) can indicate movements of the earth's crust and the location of oil.*

Evidence of movement *of the earth's crust can be found in sediments from the sea floor (below).*

The Unwilling Frontiers

The earth's polar regions have always been of compelling interest to man, but their inhospitable climates inhibit close study. Of the world's oceans and continents, those of the arctic and antarctic remain the most mysterious and least accessible.

In the early part of the twentieth century, exploration of the arctic was profoundly extended by the development of the icebreaker, the airplane, and the radio. Recently nuclear submarines capable of sub-ice traveling and the satellite have become the major modern tools for studying such areas.

The Soviet Union, with about a third of its home territory lying north of the Arctic Circle, has been one of the leaders in arctic exploration. It presently maintains a network of some 100 fixed hydrometeorological stations and geophysical observatories on remote mainland points and on arctic islands to make continuous year-round observations. In addition, automatic observatories installed on ice floes intermittently transmit data on ocean currents and temperatures,

wind direction and velocity, air temperature and pressure. The patterns of drift systems are simultaneously recorded by tracing markers. In better weather, ships and planes penetrate to more inaccessible areas to make ice and snow measurements and temperature readings, take samples from the ocean floor, and to make magnetic, astronomical, and meteorological observations.

Studies of the arctic have reversed at least one crucial theory. It had long been believed that the arctic was the prime factor in the weather in the temperate zones. It has been found, on the contrary, that the arctic, far from generating the weather in other parts of the world, is itself subject to wide changes of weather that reflect fluctuations in the general circulation of the earth's atmosphere. Contrary to what had been believed, warm air penetrates the arctic atmosphere. We now know, too, that the arctic even has frequent cyclones! Weather is not merely an arctic process but a global one. With this new knowledge, we are now far better able to make long-range weather forecasts.

In the antarctic, ice-coring has been one of the most valuable means of acquiring new knowledge. It has told us much about glaciation. From the dating of layers of sediments and ice-rafted debris, it has been discovered, for example, that that continent has been covered with ice for at least 20 million years. This finding is in sharp contrast to the estimates of five to seven million years that were previously thought accurate. It appears that a major and abrupt change in the extent of glaciation took place about five million years ago inasmuch as the ice cover then extended 200 or 300 miles farther than it does now. The melting process which resulted could have created a worldwide rise in sea level on the order of several tens of feet. Indeed, the waxing and waning of the ice sheet continues to affect sea levels all over the world. Perhaps, thousands of years from now, our coastal cities will be inundated by rising seas.

For the sake of knowledge we must pursue these forbidding environments and study them. Fundamental free research is what future scientists will rely on to understand our world and find the ways to save it.

Polar researchers explore the frozen interior of an underground cave. The secrets of such forbidding environments must be unlocked if for no more than the sake of knowledge itself.

Index

ILLUSTRATIONS AND CHARTS:

Howard Koslow—17 (top), 138-139 (top).

PHOTO CREDITS:

American Philosophical Society—16; Dr. T. D. Barnett—58 (right), 60 (top), 80, 86, 108 (bottom); The Bettmann Archive, Inc.—11, 12, 14, 22; Robert Brigham, National Marine Fisheries Service—56 (top), 58 (left); Bruce Coleman, Inc.: Oxford Scientific Films—83; Kenneth G. Compton—73 (top left); Exxon—44-45; Freelance Photographers Guild: Ron Church—101; Bruce C. Heezen and Charles D. Hollister, *The Face of the Deep,* Oxford University Press, 1971—35 (top left); William L. High —127; Hughes, Aircraft Company (from transparency supplied by NASA—114-115; Arthur C. Mathieson—128, 129 (top); Dr. John F. Michel—70; Richard C. Murphy—13, 23, 24 (bottom), 25, 103 (bottom); Musée Oceanographique de Monaco—21, 24 (top), 26-27, 29 (middle), 29 (bottom), 30-31, 40-41; NASA—78, 112-113; Naval Civil Engineering Laboratory—110 (bottom); Naval Photographic Center—42, 97, 138 (bottom), 140-141; Naval Undersea Center, San Diego—110 (top); Naval Undersea Research and Development Center—121; D. R. Nelson—59, 93; Bob Nemser—69 (bottom); The Netherlands Consulate General—71; Official Navy Photograph—96; Gerald Saunders—68-69 (top); The Sea Library: B. Campoli—46-47, 94, D. Chamberlain—102, 104, Ben Cropp—66, 67, Dr. R. Fleischer—84 (top), Henry Genthe—34 (top left), 39, 61, 89 (right), 91 (bottom), Glomar Marine—48, 51, William L. High—126, Robert Marx—119 (right), Naval Undersea Center, San Diego—111, 132 (bottom), 133 (top right), Chuck Nicklin—75, 132 (top), Ocean Systems, Inc.—62, Carl Roessler —33, 54, 55, 63, 79, 88, 129 (bottom), Scripps Institute—50, 108-109 (top), Dr. C. W. Sullivan—84 (bottom), 85, Joseph A. Thompson—36 (bottom), 122-123, 133 (bottom left), Paul Tzimoulis—73 (right); E. A. Shinn—28, 29 (top); C. Anderson Smith—73 (bottom left); Tom Stack & Associates: Ron Church—80-81 (bottom), 95, 138, Tom Myers—103 (top), Western Marine Laboratory—34 (bottom right), 57, 72, 77; Surfer Publications: John Severson—65; Taurus Photos: R. Burton—18-19, Dick Clarke—56 (bottom), 90, 100, Bob Dunn—124, Anthony Mercieca—34 (middle right), A. Conrad Newman—38, J. Pete Schroeder, D.V.M.—116, 117, Dave Woodward—133 (bottom right).